ECOWAS AND THE ECONOMIC INTEGRATION OF WEST AFRICA

ECOWAS
and the Economic Integration of West Africa

by
UKA EZENWE

ST. MARTIN'S PRESS, NEW YORK

© Uka Ezenwe, 1983
All rights reserved. For information, write:
St. Martin's Press, Inc., 175 Fifth Avenue, New York, NY 10010
Printed in Great Britain
First published in the United States of America in 1983

Library of Congress Cataloging in Publication Data

Ezenwe, Uka.
  ECOWAS and the economic integration of West Africa.

  Bibliography: p.
  Includes index.
  1. Economic Community of West African States.
2. Africa, West—Economic integration. I. Title.
II. Title: E.C.O.W.A.S. and the economic integration of West Africa.
HC1000.E93   1983   337.1'66   83-40187
ISBN 0-312-23687-5

*To my mother, Lolo,
and to the memory of my father, Iwuchukwu*

ACKNOWLEDGMENTS

I wish to thank Professor Peter Robson of the University of St Andrews, Scotland, for sharpening my interest in this area of study; Professor R.S. Bhambri and Dr Enam Ubok-Udom, both of the Department of Economics, Ahmadu Bello University, Zaria, for their useful comments; Ahmadu Bello University itself for the financial support it offered me; and of course my wife Regina and our children Uchenna, Chinwe and Chidi for their sacrifice and moral support. Finally, I am grateful to Mr Innocent Ofikwu, who typed the manuscript with speed and accuracy. Needless to say, any factual errors or misinterpretations are solely the responsibility of the author.

Ahmadu Bello University, UKA EZENWE
Zaria, Nigeria
August 1983

CONTENTS

Acknowledgements — *page* vii

Chapter

I. Introduction — 1

1. Integration: historical perspectives — 1
1(a). Background to colonial policy and integration in West Africa — 1
1(b). From pre-colonial disintegration to integration — 2
2. Decolonisation and the operation of centrifugal forces — 5
3. The definition and rationale of integration — 10
3(a). The concept — 10
3(b). The rationale — 12

II. Economic Setting and Development Perspectives in West Africa — 14

1. Resource endowment — 14
2. Basic structural characteristics — 18
3. Patterns of production and trade — 24
4. Conditions for effective integration in West Africa — 31

III. The Theory of Integration and African Economies — 42

1. The theory of integration and West African economies — 43
1(a). Reformulation and extension of the theory — 45
1(b). Limited relevance of the traditional theory to LDCs — 48
2. Role of measures of policy harmonisation — 52

IV. Distributing the Costs and Benefits of Integration — 58

1. Sources of uneven distribution of benefits — 58
1(a). Customs revenue effects — 58
1(b). Allocation of integration industries — 62
2. Institutional arrangements necessary to implement a distribution package — 64

V. The Performance of Existing Integration Schemes in West Africa — 67

1. *General* — 67
2. *The Customs Union of West African States (UDEAO)* — 72
2(a). *Structure and organisation* — 72
2(b). *Performance* — 74
2(c). *Reshaping of the UDEAO* — 75
2(d). *Impact of the UDEAO on trade* — 77
2(e). *The final collapse of the UDEAO* — 79
3. *The West African Economic Community (CEAO)* — 81
3(a). *Membership, structure and organisation* — 81
3(b). *CEAO problems* — 83
4. *West African Monetary Union (UMOA)* — 85
5. *The Council of the Entente States* — 92
5(a). *Structure and orientation* — 92
5(b). *History and activities of the Entente* — 93
5(c). *The Entente's political dimension; inter-state relations* — 99
6. *The Organisation of Senegal River States (OERS)* — 101
6(a). *Structure and organisation* — 101
6(b). *History and activities of the OERS* — 103
6(c). *Political dimension and inter-state relations* — 107
6(d). *A new organisation created: OMVS* — 109
7. *The Senegambian case* — 111
7(a). *The rationale of integration between Senegal and the Gambia* — 111
7(b). *The United Nations report* — 113
7(c). *The benefits of economic integration in Senegambia* — 114
7(d). *Recent progress towards integration* — 116
8. *Inter-State functional organisations* — 120
8(a). *The River Niger Commission* — 120
8(b). *The Lake Chad Basin Commission* — 121
8(c). *The Mano River Union* — 122
8(d). *The West African Clearing House* — 123

VI. ECOWAS: Prospects and Perspectives — 126

1. *Origin* — 126
2. *Aims* — 126
3. *Organisational structure* — 127
4. *Benefits and impact of ECOWAS* — 131

	5.	Problem areas	137
	6.	Realities and outlook	152

VII. The Role of External Forces — 155

    1. *What external forces?* — 155
    1(*a*). *The United Nations system* — 156
    1(*b*). *The multinational firms* — 161
    2. *Policy guidelines* — 167

VIII. Conclusion: Prospects for Growth — 171

    1. *Growth in the 1970s* — 171
    2. *Growth through trade* — 177
    3. *Prospects for the 1980s* — 181
    3(*a*). *Increased production of exportables* — 181
    3(*b*). *Export promotion strategy* — 182
    3(*c*). *Consolidation of regionalism* — 183
    3(*d*). *Aid* — 184
    4. *Policy recommendations* — 186
    5. *Conclusion* — 187

Epilogue: ECOWAS Policy on Population Movement and its Implications — 190

    1. *General* — 190
    2. *The Treaty and Protocol relating to Free Movement of Persons, Residence and Establishment* — 192
    3. *Implications for net importers and net exporters of labour* — 194
    4. *Policy issues and recommendations* — 196

Selected References — 199

Index — 203

MAP

ECOWAS states: main communication network — 66

TABLES

		page
II.1.	Economic Community of West African States: basic economic data	17
II.2.	Production of electrical power in West Africa, 1960, 1973 and 1978	18
II.3.	Proportional shares of mining and manufacturing activities in GDP of West African economies	19
II.4.	West Africa: destination of exports	25
II.5.	West Africa's dependence on major export categories	27
II.6.	Projected price, volume and value of primary products of export interest to West Africa, 1990	27
II.7.	West Africa: commodity composition of sub-regional imports, 1960 and 1978	29
II.8.	Openness of the economies of West Africa, 1979	30
III.1.	Static gains of economic integration among LDCs: empirical evidence	51
V.1.	West Africa: multilateral economic organisations	68
V.2.	Classification of multilateral economic organisations in West Africa	70
V.3.	UDEAO countries: relative importance of receipts from import duties and taxes	76
V.4.	BCEAO countries: net foreign exchange holdings, 1962–69	91
V.5.	Intra-Ghana Entente trade, 1968–72	96
V.6.	Mali and Senegal: trade statistics	105
V.7.	OERS member-states: estimated animal population, 1969–70	107
VI.1.	Labour supply and urbanisation in West Africa	133
VI.2.	Percentage of labour force in agriculture and the distribution of GDP	139
VI.3.	Amounts owed by ECOWAS members to the Secretariat	148
VI.4.	Subscription by ECOWAS members to the Fund's paid-up capital	149
VI.5.	ECOWAS members: dates of ratification of protocols	150
VII.1.	Some relevant UNDP inter-country projects in West Africa, 1972–76	158
VIII.1.	West Africa: growth of GDP, 1960–79	172
VIII.2.	West Africa: growth of merchandise trade	175
VIII.3.	West Africa: terms of trade, 1960–79	176
VIII.4.	West Africa: growth rates of agricultural production	177
VIII.5.	Structure of merchandise trade	179
VIII.6.	Actual and projected Sub-Saharan African debt-service ratios	185

I
INTRODUCTION

1. *Integration: historical perspectives*

Our study deals essentially with economics. But it will be useful to explore at the start how history permits us to have better understanding of the problems of economic integration in West Africa. History talks of the past; hence economic history focuses on the economics of the past. It is retrospective, whereas economics is prospective. To the extent that one is impossible without the other they speak the same language and they nourish each other.

1(a). *Background to colonial policy and integration in West Africa.* To a great extent, the present politico-economic order in Africa represents the legacies of the colonial past. All West African countries — indeed all Africa, with the exception of the minority regimes of South Africa, Namibia and the former Spanish Sahara — have achieved democratic rule and statehood, real or apparent. However, deeply ingrained habits and methods, artificial boundaries and structures, and not least ways of thinking imparted and fostered by the colonial system were not suddenly blown away by the 'wind of change'. It must be recalled that before its assumption of formal rule Western Europe had had a long contact with Africa. This is particularly true of West Africa, which was one of the first regions south of the Sahara to have any contacts with Western Europe. However, these early contacts, which were initially trade-motivated, frequently interrupted the orderly development of the people's politico-socio-economic life.

During the fourteenth and fifteenth centuries pre-colonial civilisations are known to have flourished in certain parts of West Africa, especially in the area around Benin. African blacksmiths knew how to work gold, copper, bronze and even iron. They had even developed the agricultural system of rotational bush fallow. But any orderly and natural development towards large economic units was abruptly halted by trade. The introduction of the slave trade in the sixteenth century following the territorial conquests in the Americas and the consequent opening up of rich lands to develop and exploit

---

1. See René Dumont, *False Start in Africa* (London: André Deutsch, 1966), 35. The frequent internal wars fed the slave trade, disorganised the fabric of society and blocked progress towards a political and economic order. Past achievement in this direction was dismantled.

destroyed the earliest chances of evolution and development of large economic units.[1] There is no way of gauging where the agrarian or technical African civilisation would be today if it had been able to follow a normal course of development in contact with European techniques. History cannot be remade. The most one can say, in retrospect, is that if these medieval African communities had been left alone and undisturbed to engage in legal and peaceful trade with the outside world, they might possibly have evolved into cohesive and integrated political and economic units.

1(b). *From pre-colonial disintegration to integration.* But soon the old order was changing. The years 1830-85 saw an epoch of change and revolution: it was in essence a period of transition from a predominantly slave-trading economy to one based on trade in the raw materials of the West African forest. Before this era, West Africa was the key slave market for Europe, but with the abolition of the slave trade through a complex variety of factors — political, economic, strategic, and humanitarian — Europe sought to establish legitimate trade in its place.

The buying and selling of commodities is almost always accompanied by the contact of cultures, the exchange of ideas and the mingling of peoples, and this has not infrequently led to political complications and wars. In West Africa European traders initially limited their activities to trade and avoided entanglement in African politics. Such an attitude was encouraged by the African attitude to land,[2] the hostility and suspicion of coastal chiefs, physical impediments and climatic barriers to penetration encouraged this type of attitude.

Up to the beginning of the nineteenth century, the hostility of the well-armed coastal states was a factor in preventing European invasion of their territories and of their politics.[3] By the mid-nineteenth century, however, African opposition and resistance had started to crumble in the face of a concerted European movement directed chiefly towards the invasion of the West African interior. Because the territorial struggle involved different countries — notably Britain, France, Belgium and Germany — intense national rivalries were generated. Consequently, these colonial powers decided to call

---

2. In strict West African customary law tribal land was corporately owned. The chiefs — protectors of the tribal heritage — could not sign away lands of which in reality they were merely trustees. This being the case the alienation of land to foreigners was out of the question and tribal leaders were in duty bound to oppose any encroachment on their preserves. See T.O. Elias, *Nigerian Land Law and Custom* London, 1950, 6-7.

3. See K.O. Dike, *Trade and Politics in the Niger Delta: 1830-1885*, Oxford: Clarendon Press, 1956, 1.

a conference with the ultimate objective of settling territorial claims in Africa once and for all. This was the Berlin Conference. One aim — in a sense the major one — of the Conference of 1884-5, as defined in the third 'basis',[4] was to limit the effects of future African disputes upon international relations in Europe by prescribing a new code of conduct. The final Act provided that any power acquiring territory or establishing protectorates on the coasts of Africa should at once notify all other signatory powers, and declared that possession of territory on those coasts implied a responsibility for 'the establishment of authority sufficient to protect existing rights, and, as the case may be, freedom of trade and of transit upon the conditions agreed'.[5]

Thus, having succeeded in cutting up the continent into narrow strips of territory each running from the coast into the hinterland, the colonising authorities concentrated their energies on trade, tapping raw materials and enjoying unbridled monopoly supply of manufactures.[6] Roads, railways and waterways were organised to serve these interests. No internal transport network was independently developed to connect different parts of the same territory and therefore no national or territorial economies were created as a matter of deliberate policy except of course where two adjacent territories happened to come under one flag and where such integration lessened administrative costs.

The colonial frontiers, determined by distant and sometimes ill-informed negotiators and in some cases settled by ruler and compass alone, were not well-adapted to African needs. Even ethnic divisions appear to have exerted little influence: for example, the Ewe people were divided between the Gold Coast (now Ghana) and Togo; some Yorubas live in Dahomey (now Benin), while the majority of their kinsmen are Nigerians; the Gambia and Senegal, though for long two different countries, are ethnically the same. Until the unification of the Cameroons in 1961, the south was a part of Nigeria under the British and the rest was under the French. Nevertheless, the metropolitan authorities, true to their policies, tightened their grip on these territories by means of customs, tariffs and monetary arrangements; the invisible trade of these territories was also kept largely in their hands.

4. S.E. Crowe, *The Berlin West African Conference*, London, 1942.
5. John D. Hargreaves, *Prelude to the Partition of West Africa*, London: Macmillan, 1963, 337.
6. In some instances formal administrative responsibilities were not directly assumed. Britain in 1885, for example, happily left the administration of the area beyond the Niger delta in the hands of the National African Company 'as the cheapest and most effective way' of discharging the obligations to maintain free navigation which had been accepted at Berlin. See J.D. Hargreaves, *ibid.*, 338.

It was therefore within these new borders that the technology, culture and institutions of the several colonial powers gradually made their impact during the twentieth century. Neighbouring West Africans with almost identical cultural traditions found themselves overnight subject to different languages and different doctrines in school. Meanwhile, the two major colonial powers, Britain and France, tried to achieve strong inter-territorial links within their own area of authority, albeit for political and administrative convenience. On this France was a remarkable success having pursued its favoured colonial policy of assimilation and direct rule with a missionary zeal. Indeed, in 1895 Paris took the first step to co-ordinate the activities of the individual colonial governments and to direct them towards common objectives by creating the overall government of French West Africa (AOF). The colonies were both *de facto* and *de jure* an integral part of the French union, even to the extent of having seats in the French Parliament. The eventual formation of independent governments in the colonies, even in the distant future, seemed ruled out. In effect, the economies of the colonies were integrated, especially in relation to France.

Thus the prevailing circumstances set in motion other forces making for greater integration and unification. Within the French or British territories, trade and migration took no account of territorial boundaries. The spill-over effect of education also contributed in no small measure to reinforce this trend towards colonial economic integration. Institutions of higher learning were established only in a few strategic centres and they drew recruits from all parts of the dependencies. For instance, there was only one university (at Dakar) for the whole of French Africa from the end of the Second World War up to the period of independence. British West Africa, though marginally better off, had no more than three universities (Ibadan in Nigeria, Legon in Ghana and Fourah Bay in Sierra Leone) by the close of the 1950s. Furthermore, there was the unifying influence and the spirit of belonging which the adoption of a common language in education created — although this has also created and fostered the polarisation between the Francophone and Anglophone parts of West Africa which to date remains an important divisive element.

In other aspects of economic life, effective colonial integration was in vogue. As already noted the monetary systems in colonial West Africa were centralised. French West Africa had a common currency (which has survived),[7] a range of common services and a

---

7. The West African Customs Union (UMAO), which includes all the francophone West African countries except Guinea and Mali, uses a common currency, the CFA

customs union which served as a mechanism through which resources were distributed from the wealthier coastal states of Senegal and Ivory Coast to their less well-endowed peripheral inland neighbours. Even in British West Africa, which was less closely tied to the metropolis, there was a common currency issued by the West African Currency Board. The establishment of research institutes and the organisation of common services and marketing boards[8] was also on an inter-territorial basis.

What therefore follows from the foregoing account can be summarised as follows:
(i) pre-colonial West Africa saw a measure of civilisation, and by the nineteenth century there was an evident trend towards the construction of larger and more effectively centralised communities, which the process of colonial penetration interrupted;
(ii) the colonial boundaries, arbitrary and imposed, were far more rigid than the pre-colonial ones;
(iii) the colonial authorities, each within its own area of jurisdiction, often achieved strong inter-territorial ties through the application of politico-economic integration policies, essentially for pragmatic reasons of administrative economy and convenience.

2. *Decolonisation and the operation of centrifugal forces*

Given the level of economic integration achieved in West Africa before independence by each colonising power within its area of authority, one might have expected unimpeded progress, even if at a slower pace, towards closer economic co-operation after the achievement of self-rule. But this has not happened. Even on a continental scale, the story of economic integration between African states since the end of colonial rule has been, with few exceptions, the story of disintegration rather than integration. In spite of the present state of affairs, there is almost a staggering unanimity in principle among African governments, as exemplified in speeches, conferences and resolutions, on the urgent need for some form of close economic co-operation. However, that has been shadow rather than substance.

---

franc and has a common central bank (BCEAO). Following a co-operation agreement in 1962, France guarantees convertibility of the CFA franc into the French franc.

8. Before 1947, the West African Produce Control Board handled the marketing and management of British West Africa's agricultural products. Thereafter, however, national Marketing Boards, which were set up, took over and shared out assets of the West African Produce Control Board. See G.K. Helleiner, 'The Fiscal Role of the Marketing Boards in Nigerian Economic Development 1947–61' in E.H. Whetham *et al.* (eds), *Readings in the Applied Economics of Africa*, Cambridge University Press, 1967.

One may now ask what are the forces that have encouraged disintegration in West Africa? In particular, can the identification of these obstacles to integration lead to their removal? A cursory survey of the recent history of integration in West Africa, which has been discussed above, will help to identify some of the problems.

One of the most important factors here seems to centre around the question of national sovereignty. The cadre of African leaders who fought for independence did so on the basis of the nation-state and the right of each country to govern itself. Hence there was a strong feeling among the African nationalists that political and economic salvation lay within the independent nation state. Kwame Nkrumah's popular slogan 'Seek ye first the political kingdom' seems to have reflected the mood of most of the African leadership on the eve of independence. Because integration involves a conscious surrender of a measure of national sovereignty in policy formulation and execution, or at least a pooling of it, it was therefore not hard to see how the consciousness of newly-won independence exerted strong influence on the general approach to economic integration. Indeed, aside from the issue of whether economic integration holds out the promise of improving the viability and performance of each territorial economy, the key question for the countries under examination is that of reconciling themselves to the (real or apparent) economic sacrifice of part or all of its sovereignty — the degree of sacrifice being a function of the kind of market arrangement. Thus the present weakening of economic ties is justified on the grounds that political power was transferred at independence to the territorial, not to the supranational units. Thus maintaining the *status quo* after independence was a natural corollary.

Furthermore, the post-independence trend towards disintegration has been rationalised on the grounds that unity under colonial rule was illusory. As Hazlewood succinctly put it, 'it was a unity imposed from outside for the administrative convenience of the colonial power — it was a unity of Europe in Africa, reflecting the hegemony of the metropolitan country over its various colonies. It was not to be expected that, with the removal of Europe from the scene, the unity would necessarily continue.'[9] Although this is generally true, there is another side to the coin. In the case of French West Africa, for example, the sequence of measures introduced by the French government long before independence had a destructive effect on the integration that had been established between the French colonies.

---

9. Arthur Hazlewood (ed.), *African Integration and Disintegration*, Oxford University Press, 1967, 3.

The French Union was created under the 1946 constitution of the French Republic which established territorial councils in the Federation of French West Africa (AOF) as well as in French Equatorial Africa (AEF). The year 1952 saw the transformation of these territorial councils into Assemblies of a political nature. This process was carried further with the establishment of universal suffrage under the *loi-cadre* of 1956, and a redistribution of power in favour of the territories. Two years later the territories were asked to choose between three alternatives, namely complete independence, independence within the French community and absorption as a *Département* of France. Only Guinea opted for complete independence; all the others chose independence within the community and were granted full independence by a stroke of the pen in 1960. It cannot therefore be persuasively argued that independence was the only cause of disintegration. It was more an effect than a cause.

Colonial critics tend to see in this chain of events leading up to independence as the 'colonial plot' or what Hodgkin calls 'false decolonisation'.[10] They argue that the process of disaggregation and fragmentation prior to independence was a tactical move clearly in line with the general interests of the Western powers. According to this view the present politico-geographic map of Africa, as a mosaic of petty states which are extremely vulnerable to external pressures, preserves the basic relationship of Western dominance and African dependence by other means, after the transfer of formal political power.[11]

Again, the above statement seems to tell only half the story. It must be emphasised that the pressure from African nationalists for self-rule was an irresistible force and that the level of political awareness varied from one territory to another, with the result that different territories were ready for independence at different times. To this extent it could be said that the colonial authorities were motivated by the desire to ensure that independence was handed over to the right people at the right time and that they merely responded to this urge.

The doubts surrounding the net benefits from integration in less developed countries (LDCs) bring us to the second point. Despite the *a priori* gains arising from market integration, it can be contended that, as a strategy of economic development, customs unions do not

---

10. See Foreword by T. Hodgkin in R.H. Green and Associate, *Unity or Poverty?*, Harmondsworth: Penguin Books, 1968, 14.
11. *Ibid*. Even the encouragement of such groupings and associations as the West African Customs Union, the Conseil de L'Entente, OCAM or the EEC-Associateship by the West is regarded as not inconsistent with the neo-colonialist theory so long as their objectives are limited and attached to safe Western leading-strings.

yield great net benefits to the joint economic area in the early years. While the quantification of net gains might be difficult and even contentious, it would seem that, generally, the net contributions to the economic development of Africa from market integration in the foreseeable future are likely to be marginal. Green makes the bold generalisation that 'it seems highly unlikely that the net present benefit from any existing structure of African economic integration exceeds 1-2 per cent of domestic products.[12] This no doubt stems, *inter alia,* from the structural imbalance of the economies of Africa which limits the volume of tradeable goods among African states. It may well be that, despite the large allocation of top-level political, civil service, academic and research time now being directed to market integration studies in LDCs, the enchantment generated by the subject has only been equalled by the degree of caution with which African countries have actually approached it.

The question of marginal net gains is often exacerbated by the problem of probable or real uneven distribution. For while the logic of specialisation and of economies of scale centres on the general increase of welfare of the members of a free-trade community, the theory has no universally acceptable mechanism for ensuring the equitable distribution of the welfare gains among the individual members. There are, however, at least two common distribution devices for tackling this problem: the mechanism of fiscal transfers and the so-called 'managed' specialisation which details in form of a legislative instrument the formulas for equitable distribution of benefits. The trouble with these tools is that they tend to offer less than what the weaker economies want — 'revocable subsidies' and licences for some new regionally-conceived industries instead of strategic growth points, although the former, under certain circumstances, could create linkages comparable to the latter. At the same time they hold back the richer countries which can only make limited sacrifices at the present stage of their development.

It was on this issue of unsatisfactory distribution strategy that the West African Customs Union virtually disintegrated. Ivory Coast, with about half of the region's exports and tax revenue but only a quarter of its imports and total expenditures, was unwilling to continue subsidising the weaker economies, especially that of

---

12. R.H. Green and K.G. Krishna, *Economic Co-operation in Africa: Retrospect and Prospect,* Oxford University Press, 1967, 29. See also P. Robson, *Economic Integration in Africa,* London: George Allen and Unwin, 1968, 91. Professor Robson in quantifying the contribution of integration in the economic development of the three members of the defunct East African Economic Community (Kenya, Tanzania and Uganda) found that in 1965-70 the gains from integration would only imply an improvement in the regions annual growth of GDP of 0.5 per cent over the period.

Senegal. It sought rapid industrial expansion and markets for its industrial sector. It therefore organised the Entente, which excludes Senegal. Even so, the less developed Entente members wish to have leeway in promoting a number of consumer goods industries to replace imports from their richer neighbours. Inability to realise this aspiration increased disillusionment and the likelihood of withdrawal. The East African Community faced the threat of disintegration between 1964 and 1965 for similar reasons, but in the end wise counsel prevailed and the reshaped community under the 1967 Treaty tried to minimise the sacrifices of individual members. When eventually the Community collapsed, it was partly on this score. However, the wishes of individual countries apart, the institution of checks and balances in a common market cannot be avoided if it is to survive to the mutual advantage of all.

Finally, another factor that might have affected the attitude of post-independence Africa towards integration is the problem of institutional framework building. The establishment of a competent administrative body, flexible and devoted, whose function would be to ensure the equitable distribution of gains from customs union and to safeguard sometimes conflicting national interests is not very easy in the present technical conditions of Africa.

There is also a political aspect to the matter. Let us consider the EAEC case. Under the 1967 Treaty, the three heads of state assisted by ministers constitute the principal executive authority of the community; all bills and annual budgets must receive the assent of all the three heads of state before they can be executed. This implies a considerable range of mutual and political understanding between the three, especially in their foreign policies in order to avert disruptive inter-state hostilities. In fact, the 1971 coup in Uganda and the refusal of President Nyerere of Tanzania to recognise the regime of Idi Amin, which in turn held up the passage of the 1971/2 annual budget, once more underlines the need for political understanding among common market countries.

Needless to say, in one significant sense economic integration is in itself a political issue; it is an aspect of the global strategy of economic development which is essentially political. Thus the political or ideological philosophies of the new states of Africa are likely, directly or indirectly, to have some effect on their long-term economic policies.

On the whole, none of these barriers to integration is insuperable, given a genuine and strong desire to overcome them and political goodwill.

3. The definition and rationale of integration

3(*a*). *The concept.* 'Integration' in ordinary usage means unification or putting parts together into a whole. In economic literature the term 'economic integration' is sometimes hard to pin down to a precise definition. Some authors include social integration in their conceptualisation; others define integration from static or dynamic standpoints. From the static point of view integration is considered as a state of affairs which would obtain at the end of a fairly long process leading to the complete merger of national identities. The dynamic view, on the other hand, sees integration as a process whereby discriminations existing along national borders are progressively removed between two or more countries. Even further afield there are other definitions which view integration as the mere existence of some measure of trade relations between independent national economies. Integration, according to this view, progresses in stages from its lowest to its highest forms: the freeing of barriers to trade ('trade integration'), the liberalisation of factor movements ('factor integration'), the harmonisation of national economic policies ('policy integration') and finally the complete unification of these policies ('total integration').[13]

For the purpose of this study we intend to adopt the dynamic concept of integration — although in the long run the dividing line in the static-dynamic dichotomy is blurred. In this sense, our operational definition of integration is the gradual but steady process of harmonised tariff disarmament along with the removal of other barriers to trade between the contracting parties to their mutual advantage. This seems to us to be the most feasible and achievable objective of market integration in today's West Africa. A workable pattern of integration has to be realistic and pragmatic. It should neither embrace the doctrinaire attitude of the 1960s nor demand political union as a pre-condition for economic integration.[14]

Indeed, the first stage of integration which is much closer to economic co-operation — a process of stage-by-stage lowering of tariff and other barriers to trade — seems a convenient starting-

---

13. Bela Balassa, 'Types of Economic Integration' in Machlup (ed.), *Economic Integration, Worldwide, Regional Sectoral*, London: Macmillan Press, 1976, 18.
14. Of course, economic integration cannot operate in a political vacuum; a good many integration movements could have been primarily politically inspired. Also apart from the interdependence between politics and economics, economic integration is not an end in itself; it is a means to an end and the end is essentially a political goal. Even so, while a successful integration would require effective political understanding and goodwill among its members, complete political unity need not be a *sine qua non* especially in the early phases of economic integration.

point, politically and otherwise, for emergent LDCs. But when and if there is an uninterrupted progress towards the completion of the integration process, then the static concept could be applied. So in the West African context we are primarily concerned with the 'integration path' which ultimately leads to the 'integration goal' when everything would be static and 'dead' and not the other way round.

Furthermore, there is one more conceptual aspect of integration which should be put in its proper perspective. The economic significance of national borders is that they introduce discontinuities in the flows of commodities and factors of production. And it is these discontinuities which actually lead to effective discrimination in the economic sphere. But when the discriminatory tariff walls are dismantled obstacles to intra-zonal trade will be lessened or even completely removed. However the degree of free movement of goods and factors would be a function of the stage of integration. In classifying the stages of integration, five major categorisation are usually made. They include:

(i) the free trade area, which implies the removal of quantitative restrictions and customs tariffs;
(ii) the customs union, which unifies the tariff of the countries within the area against outsiders;
(iii) the common market, where all restrictions on factor movements within the area are abolished;
(iv) the economic union, where economic, monetary, fiscal, social and counter-cyclical policies are to some extent harmonised;
(v) the supranational union, where the respective governments completely abandon their sovereignty over the policies listed above and a supranational authority issues binding decisions.

Like most classifications, these are somewhat arbitary. Some forms of market integration may well fall within these categories. The first three stages concern mainly trade and factor integration but more often than not such measures require a complementary payments arrangement to make them work. And beginners in the field of integration usually start somewhere between the first two, with or without some elements of the third category. West Africa falls squarely within the beginners' class. The last two stages, which are very advanced, would in any case have very little chance of success in most LDCs.

Although the forms of integration mentioned above represent varying degrees of economic integration, they nevertheless share two basic characteristics. First, they promote expanded intra-zonal specialisation and exchange through the reduction or elimination of trade restrictions among the union members; and secondly, they entail discrimination of one kind or another against non-member-countries.

3(b). *The rationale.* The driving force behind the widespread interest in economic integration in Africa is twofold. The first motive is political. As noted earlier, colonialism in Africa left behind it a geopolitical configuration of divisions and fragmentations. Many of the new African states, although nominally independent, were so small and weak both politically and economically that they had little prospect of rapid economic development on their own. This made them extremely vulnerable to external pressures, which worked to perpetuate African dependence upon foreign powers. There was therefore a widespread feeling in Africa during the 1960s to free the continent from its external dependence and to provide the safeguards and benefits of interdependence through the achievement of economic integration of one kind or another.[15]

The second and much more fundamental reason is economic. Given the micro-states and the export-oriented, lop-sided, poor economic structures inherited from the colonial regimes, which were in need of reconstruction, integration was seen as a means of helping to overcome the disadvantages of small size and of making possible a greater rate of balanced economic growth and development. As the UN Committee for Development Planning put it, 'The creation of unified multinational markets would make possible faster expansion and greater economic diversification of the African economy and particularly of its industry. It would also enhance productive efficiency by permitting increased specialization and the operation of industries on a more economic scale. Further, it would help in overcoming barriers to development appearing in foreign trade with the developed parts of the world.'[16]

To this one might add that wider markets, which integration is bound to bring about, would, other things being equal, attract more foreign capital and create more employment opportunities.

Thus it could be said, in a nutshell, that the key rationale of economic integration in Africa today is the acceleration of balanced growth in the partner countries — either in the short or in the long run. Therefore, freeing trade or factor movements is not an end in

---

15. At the more economic level, regional economic co-operation including intergovernmental organisations, on which the ECA has been on the vanguard, has been encouraged. But on the more ideological plane, there was the Pan-African movement which had as its goal continental government of a united Africa. Achievement on both accounts however have been minimal.
16. United Nations, *Economic Co-operation and Integration in Africa: Three Case Studies*, New York, 1969, (ST/ECA/09), iii.

run. Therefore, freeing trade or factor movements is not an end in itself but a means to reach higher levels of output. Indeed, the benefits of the integration process have to be judged by the criterion of whether, on balance, the area's growth rate is faster than it would otherwise be or not. As has already been indicated, there is little doubt that the marginal growth contribution which integration holds out for LDCs in the foreseeable future underlines the cautious optimism with which most African governments have so far approached economic integration issues.

Of course, everything depends on the future of the industrial sector in Africa. Ironically, one of the major objective functions of central planners in LDCs is the expansion of the industrial sector, a policy which would have suggested a much more positive attitude to integration than appears to be the case on the African continent at present.

Since the mid-1950s the formation of economic groupings, especially among LDCs, has become a common phenomenon. But the experiences of many of these groupings, including those of West Africa, have been very disappointing. This book discusses the historical development, experience and lessons of economic co-operation in West Africa. It tries to highlight the causes of the failure of the defunct, and the problem areas of the extant, integration schemes in the sub-region with a view to offering policy suggestions for future action.

II
ECONOMIC SETTING AND DEVELOPMENT PERSPECTIVES IN WEST AFRICA

The introductory chapter sought to develop a historical link to this study, which in the present chapter focuses on the economic background of the West African sub-region. To acquire, at the outset, a realistic appreciation of the opportunities, problems and prospects of economic integration in the sub-region one has to examine the economic setting, development perspectives and recent economic and political developments in the area which have immediate bearings on economic co-operation.

1. *Resource endowment*

In terms of mere physical size, West Africa is a vast region, and the sixteen countries[1] of the sub-region have a combined population of about 145 million spread over an area of 6 million square km. (2,380,000 square miles). Despite the size of the population — about one-third of all Africa — the average density of the region is a mere twenty-four persons per square kilometre. Even so the mean figure conceals the wide variations in local and national densities.[2] On the international scale the population density varies from one person per sq. km. in Mauritania to eighty-nine persons in Nigeria (Table II.1).

Even with adjacent territories there are great contrasts. Compare in this respect Niger's density of four persons per sq. km. with the figure for Nigeria. To a lesser extent the same is true of Liberia with its sixteen persons per sq. km. compared to Sierra Leone which has

---

1. The sixteen countries of the sub-region, all of which are ECOWAS members, are Benin, Cape Verde, Gambia, Ghana, Guinea, Guinea-Bissau, Ivory Coast, Liberia, Mali, Mauritania, Niger, Nigeria, Senegal, Sierra Leone, Togo and Upper Volta.

2. The three land-locked countries, Mali, Niger and Upper Volta, together with Mauritania, account for over 60 per cent of the total area but contain only 12 per cent of the population. It may be noted that the 14 countries of East Africa which have the same size (6 million sq. km.) as West Africa and a combined population of 70 million have lower densities still. Of course, Africa as a whole is the most sparsely populated region in the world, apart from Antarcica and Oceania. With only 9 per cent of the world's population Africa has about a quarter of its land surface. Thus pockets of population concentrations derive largely from history and sociology rather than from the stark land/man ratio that presses on most of Asia today.

Resource endowment

Table II.1
ECONOMIC COMMUNITY OF WEST AFRICAN STATES: BASIC ECONOMIC DATA

(1)	Population 1979 (millions) (2)	Population growth rate 1960–79 (average) (3)	GNP 1979 prices (US $m.) (4)	GNP per capita 1979 (US $) (5)	Growth rate of GDP at factor cost 1960–75 (average) (6)	Total land area (1,000 sq. km.) (7)	Density 1979 (persons/sq. km.) (8)	Manufacturing & electricity growth rate 1960–75 (average) (9)	Trade as % of GNP 1976 (10)	Intra-ECOWAS to world total 1968–72 (average) (11)	Purchasing power density (1979) (12)	Share of inter-ECOWAS trade 1968–72 (average %) (13)
Benin	3.4	2.7	850	250	2.0	113	30	8.0	42.4	15.0	5,132	0.20
Cape Verde	0.3	2.9	78	260	0.9	n.a.	n.a.	8.8	n.a.	n.a.	n.a.	n.a.
Gambia	0.6	2.2	150	250	5.0	10	60	4.0	110.0	5.0	15,000	0.02
Ghana	11.3	2.6	4,520	400	2.0	239	47	7.5	35.9	1.2	18,912	0.40
Guinea-Bissau	0.8	–0.4	136	170	1.0	n.a.	n.a.	9.8	n.a.	n.a.	n.a.	n.a.
Guinea	5.3	2.8	1,484	280	2.0	246	22	4.0	25.5	n.a.	6,032	0.04
Ivory Coast	8.2	3.6	8,528	1,040	7.5	322	25	10.8	24.0	5.0	26,484	0.60
Liberia	1.8	2.3	900	500	3.5	111	16	3.5	116.9	2.0	8,108	0.08
Mali	6.8	2.2	952	140	1.0	1,202	6	6.0	37.7	23.8	792	0.30
Mauritania	1.6	2.1	512	320	5.0	1,806	1	7.5	85.5	0.3	283	0.20
Niger	5.2	2.7	1,404	270	3.5	1,267	4	10.1	21.6	22.0	1,108	0.20
Nigeria	82.6	2.5	55,342	670	6.0	924	89	8.0	62.1	1.7	59,844	0.40
Senegal	5.5	2.6	2,365	430	0.8	196	28	3.5	52.7	20.0	12,066	0.70
Sierra Leone	3.4	2.2	850	250	4.7	72	47	6.5	53.9	1.0	11,806	0.09
Togo	2.4	2.7	840	350	5.8	57	42	10.7	57.9	3.7	14,737	0.09
Upper Volta	5.6	2.0	1,008	180	3.0	274	20	3.0	22.8	54.7	3,679	0.30
Total	144.8		79,919	(av.) 500								ca. 3.62

Notes: 1. In column 10 dependence on trade or measure of openness of an economy is given as the ratio of trade (exports plus imports) to GNP.
2. Purchasing Power Density is defined as GNP per unit of area.

Sources: *World Bank Atlas* (Washington DC, USA, 1976); *Africa: South of the Sahara 1975*; and United Nations, *Handbook of International Trade and Development Statistics 1972*; EEC, *The Courier*, 31, March 1975; World Bank, *World Development Report*, 1981.

over 47 persons per sq. km. Within national borders the general characteristic feature of uneven population distribution pervades. Consider that while Northern Nigeria has few persons per sq. km., the Ibo heartland of the south-east has one of the highest densities on the African continent.

Partly because of the lack of water in many parts of the region, and partly because of the over-concentration of development projects in the southern urban centres, which have attracted rural drifters, one can find extremely high population densities in some areas of the south and very low ones in the north. A second reason, which centres more on historical and sociological than geographic and economic factors, can be adduced to explain the irrational pattern of population distribution in West Africa. The era of the slave trade and inter-tribal warfare which it inspired must have resulted in the depopulation of large areas and in the development of high concentrations of population in other relatively inaccessible districts. This tendency towards uneven population distribution was reinforced by later patterns of settlement.

Although the establishment of the most important population concentrations took place before the colonial conquest, the institution of common tribal ownership of land acted as a brake on large-scale migrations of people to other less densely populated tribal lands without recourse to war. When, following the consolidation of colonial conquest, tribal warfare finally disappeared, the arbitrary sharing out of the region by different European nations and the introduction of different national laws and system with respect to trade and movement of labour doomed to failure any hope of a future mass redistribution of population through migration. Hence this problem remains and, as we shall see later in this study, the question (together with its implication) of migrant workers from the landlocked to the coastal countries has important bearings on economic co-operation in West Africa.

In terms of agricultural resources, the vast span of the sub-region provides arable land of varying fertility, ranging from the rich rain forest belt in the south to the poor and dry semi-desert in the north. Lying within longitudes 20° W. and 15° E., and latitudes 17° N. and 10° S., the most striking feature of the sub-region is the great diversity of its climatic conditions and topography. It stretches some 2,880 km. from the rain forest region on the coast to the land-locked countries on the edge of the Sahara. There is hardly any generally accepted geographical definition of West Africa. The region appears neatly separated, but some of the southern and eastern boundaries are less precise. However, for the purpose of this study the ECA definition referred to above is adopted.³

In the areas of primary agricultural production, West Africa contributes about 72 per cent of the cocoa, over 76 per cent of the volume of palm kernels and over 32 per cent of the volume of palm oil entering world markets. The bulk of the groundnuts entering international trade come from two countries in West Africa, Nigeria and Senegal. Other export products of the sub-region include: coffee, cotton, hides and skin, phosphates, tobacco, rubber, timber, plywood, sesame, bananas, pineapples and kola nuts.

The endowment in other natural resources is equally impressive. Vast reserves of electric resources are present in certain parts of the region (Table II.2). Nigeria produces both coal and petroleum. In fact, the Nigerian oil industry enjoyed a boom throughout the 1970s, and it is still the backbone of the economy. Production grew rapidly immediately after the civil war, reaching an all-time record of 823.3 million barrels of crude in 1974. In that year 92.6 per cent of total export earnings came from oil, which also accounted for over 85 per cent of federally collected revenue. Although since 1975 output has tended to fluctuate, the dominant position of the oil sector in the economy remains unimpaired and is expected to continue till the end of the 1980s.⁴

Other minerals are also found in West Africa. Over 60 per cent of world production of gold derive from Africa, the bulk (excluding South Africa) from West Africa; 96 per cent of world production of diamonds come from Africa, the bulk (outside the Congo) from West Africa. Over 45 per cent of world output of manganese is produced in Africa with West Africa accounting for 34 per cent, and West Africa produces over 27 per cent of the African output of iron ore. There are also reasonable quantities of bauxite. Extensive deposits of other minerals in the region have been reported, but their extent is not yet determinable.⁵ Thus it follows that the poor record of integration schemes in West Africa cannot seriously be blamed only on lack of natural resources. Obviously many of the problems are man-made.

---

3. In a way to describe West Africa as a 'region' is merely to use a geographical expression, and should not give the impression that it is already a single 'international economic region' or 'market area'. The area has never been an integrated market, although with the formation of ECOWAS, the prospects are brighter.
4. See Central Bank of Nigeria, *Economic and Financial Review*, December 1977.
5. See R.H. Green and Seidman, *Unity or Poverty?: The Economics of Pan-Africanism*, Harmondsworth: Penguin Books 1968, 53. Also see N.A. Cox-George in *The Nigerian Journal of Economic and Social Studies*, 5, 1, March 1963.

18   *Economic setting and development perspectives*

Table II.2
PRODUCTION OF ELECTRICAL POWER IN WEST AFRICA, 1960, 1973 AND 1978

	Total installed capacity (1,000 kw.)			Installed capacity per 1,000		
	1960	1973	1978	1960	1973	1978
Benin	6	14	15	3	5	4
Cape Verde	1	6	6	5	21	19
Gambia	4	10	8	10	20	14
Ghana	103	892	900	15	95	83
Guinea	n.a.	150	175	n.a.	36	37
Guinea-Bissau	2	8	8	4	16	11
Ivory Coast	32	357	502	10	77	95
Liberia	22	298	300	22	183	163
Mali	n.a.	27	43	n.a.	5	7
Mauritania	n.a.	39	40	n.a.	31	29
Niger	3	18	20	1	4	4
Nigeria	173	855	960	4	14	14
Senegal	56	120	130	18	28	27
Sierra Leone	21	87	95	10	33	29
Togo	n.a.	21	24	n.a.	10	10
Upper Volta	4	16	20	1	3	3

Source: UNCTAD, *Handbook of International Trade and Development Statistics*, 1980, 372–5.

2. *Basic structural characteristics*

In general, the level of development of any region is to some degree a function of the structure of its economy. And the level of development itself is reflected in the structure of the economy. West Africa, in the socio-economic sense, is typically an underdeveloped area, exhibiting the common characteristics of underdevelopment. Although there may be many of these characteristics, we will touch upon the four most important ones upon which others hinge in the matter of integration.

First, the principal feature of the economic structure of Africa is the predominance of agriculture. West Africa is no exception. The average share of agriculture, including livestock, forestry and fishing, in the total GDP of the subregion is between 50 and 55 per cent,[6] although this average figure covers important variations among the countries. However, in three of them (Liberia, Nigeria and Senegal) the contribution of agriculture is only of the order of 30

---

6. UNECA, *Economic Co-operation and Integration: Three Case Studies, 1969* (ST. ECA/109), 50.

per cent. The reason for this below average figure is not far to seek. Extractive activities are very important in Liberia and Nigeria, and Senegal has one of the most highly developed industrial sectors of West Africa (see Table II.3). A second reason is that agriculture in Senegal is centred mainly on groundnut production, which is one of the victims of natural substitutes in the world market. The government's agriculture diversification programme, whatever its achievements, has not substantially altered the picture.

A further general feature of the agricultural sector, aside from its predominant position, relates to the dichotomy between the non-monetised or subsistence and monetised sub-sectors. In all the countries under discussion the non-monetised subsector accounts on the average for more than half of agricultural production, and for considerably more than half in most other LDCs. Compared with subsistence agriculture, the monetised sub-sector is usually smaller and is mainly oriented, directly or indirectly, towards exports. This is of course a reflection of the general colonial policy on land and trade in West Africa during the first half of the twentieth century, whereby the natives were actively encouraged to grow cash and food crops side by side. The former promoted export trade while the latter was meant for subsistence.

Table II.3
PROPORTIONAL SHARES OF MINING AND MANUFACTURING ACTIVITIES IN GDP OF WEST AFRICAN ECONOMIES (%)

	Manufacturing			*Mining*
	1960	1977	1979	1977
Benin	3	n.a.	8	6.8
Cape Verde	n.a.	1.8	n.a.	0.3
Gambia	n.a.	n.a.	n.a.	1.1
Ghana	10	13.1	n.a.	1.3
Guinea	n.a.	6.3	5	18
Guinea-Bissau	n.a.	1.0	n.a.	0.0
Ivory Coast	7	10.7	12	0.2
Liberia	n.a.	8.8	6	22.8
Mali	5	n.a.	6	12.6
Mauritania	n.a.	4.7	8	17.1
Niger	4	5.2	10	8.4
Nigeria	5	5.4	6	28.4
Senegal	12	16.5	19	2.5
Sierra Leone	n.a.	7.4	5	10.2
Togo	8	n.a.	7	11.9
Upper Volta	8	9.5	14	0.1

Sources: World Bank, *World Bank Report*, 1981 and UNCTAD, *Handbook of International Trade and Development Statistics, 1980.*

The second important characteristic of the West African economies is the weakness of the industrial sector. This surely, follows from the first point. With the notable exception of Senegal, the share of the manufacturing sector in total production of West African countries in generally below 15 per cent (see Table II.3). There is evidently a considerable variation in the share of this sector in the GDP among individual countries. Ghana and Togo started well in the 1960s; but the manufacturing sector has slowed down or deteriorated in both countries ever since. On the other hand, there are other countries — notably Ivory Coast, Niger and Upper Volta — whose manufacturing sectors have done relatively well. In the absolute sense, Nigeria has probably one of the most advanced sectors in the region, but the ratio of manufacturing output to GDP has been diminished by the emergence and dominance of the oil sector.

Manufacturing has one notable characteristic in this region: it depends to a great extent on agricultural production.[7] In other words, the structure of manufacturing output is heavily biased towards the processing of agricultural raw materials.

The food, beverages and tobacco component of output accounts for between one-third and half of total manufactured output. This consists mainly of the production of beer, soft drinks, cigarettes, crushed vegetable oils and a few other items such as bakery and flour milling. Textile, clothing and footwear constitute the second most important component of manufacturing production. In many West African countries, chemicals and chemical products are also significant, accounting for between 10 and 20 per cent of total output. Even so, this component reflects the existence of oil refineries in the area rather than the existence of integrated petrochemical or chemical industries as such. With the exception of cement industries, other industry groups are usually not significant. The structure of manufacturing production described above reflects the relatively low level of economic development in the region, as well as the undiversified nature of these economies, with the exception of Ghana, Nigeria and Ivory Coast.

The studies of Kuznets (1966) and Chenery and Syrquin (1975) reveal that the transformation of the structure of production is at the heart of the economic and social changes that characterise economic development. They point out that 'not only is there a strong statis-

---

7. This is probably the effect of the policy of industrialisation based on import-substitution. Historically, this form of industrialisation had always started off with the processing of agricultural raw materials and simple manufactures essentially to replace imports, but quite often with a margin for exports. See Arthur Lewis, *Industrialization in Ghana*, 1953.

tical association between the rise of industry and the level of *per capita* income, but virtually all countries that have achieved high living standards by any measure have also industrialised to a substantial degree.'[8]

Within the framework of analysis established by these writers, it appears that West African countries can safely be classified as being in the first stage of industrial development. But West Africa needs to move to the second stage in order to produce intermediate and capital goods and consumer durables. However, the critical move from the first stage with its emphasis on the processing of agricultural raw materials and simple manufactures to the second stage of industrialisation requires a wide market which individual African countries do not possess. Herein lies the case for development planning on a sub-regional or co-operation basis. Indeed, for the individual micro-state of West Africa, economic integration is a *sine qua non* for industrial development.

A third characteristic of West African economies is the highly limited forward and backward linkage effects of the manufacturing sector. The analysis of the structure of output above reveals that most of the output of the manufacturing activities is composed of final consumer goods, hence the sales of output to other sectors for use as input is necessarily very small. Thus the forward linkage effect of manufacturing in West Africa is generally low. The same can be said of the backward linkage effects of manufacturing in the region; most of the region's industries, whether consumer or capital goods industries, can be properly described as assembly industries. The beverage industry, for instance, imports not only the contents of the bottles but also the bottles themselves. Even in the textile industry in many countries, the raw materials are imported. In the few capital goods industries that exist, virtually all components are imported and merely assembled into the final product. Consequently, purchases of inputs from local producers for domestic manufacturing are quite limited in extent.

Conceivably, the enclave nature of manufacturing output in the sub-region can be substantially minimised if industrial projects which have higher backward and forward linkages are selected. And given the importance of market size as a pre-requisite for the fuller realisation of the linkage effects of manufacturing activities, meaningful inter-industrial linkages can only be achieved through

---

8. See S. Kuznets, *Modern Economic Growth*, New Haven: Yale University Press, 1966; H. Chenery and M. Syrquin, *Patterns of Development: 1950–70*, Oxford University Press, 1975. Also see H. Chenery and M. Syrquin, 'A Comparative Analysis of Industrial Growth', paper presented at the 5th World Congress of the International Economic Association, Tokyo, 1977.

concerted planning within the framework of co-operation.

A fourth but more recent economic feature of West Africa derives from the level of under-utilisation of resources. In the conventional but 'unsettled' theory of Development Economics, LDCs are generally categorised as 'regions of disguised unemployment and underemployment', and both terms were meant to apply strictly to the traditional sector. But as further enquiries into these concepts[9] have shown, under-utilisation of productive resources in West Africa, as indeed in most other LDCs, is not peculiar to the agricultural sector: it is now visible and undisguised in the modern sector as well. The key reason for this is deeply embedded in the sheer lack of equilibrium between the supply of, and demand for, labour.

The excess supply of labour could be blamed on a set of socio-economic variables: first, the general rural-urban wage differentials offer an added incentive to rural drifters; secondly, the rapid acceleration of schooling reinforced by high population growth in the countryside has speeded up the drift of young people to the towns; thirdly, development and welfare expenditure have been concentrated disproportionately in towns; and finally 'capital deepening', in itself the by-product of derived development and borrowed automated technology, has tended to create some sort of technological unemployment. Disparities in the levels of unemployment exist in the region but every country surely has its own growing pockets of unemployed and underemployed people.[10]

On the demand side, a number of factors making for the sluggishness of labour demand come readily to mind. Nearly always the government is the major employment sector partly because of its relative size and partly because the sluggishness of the private sector forces it to go on expanding, sometimes into areas more suited to the private sector. But the government's ability to go on expanding is limited by its resources. Given the weakness of the indigenous private sector *vis-à-vis* the government to generate more employment, industrialisation is often carried out by private foreign enterprises attracted by elaborate government incentives. But it would be wrong to pretend that the foreign investor and the government will

---

9. For discussion on this subject see A. Lewis, 'Economic Development with Unlimited Supplies of Labour', *The Manchester School*, May 1954; G. Ranis and J.C. Fei, 'A Theory of Economic Development', *American Economic Review*, Sept. 1961, 51, 533–58 and H.T. Oshima, 'The Ranis-Fei Model of Economic Development', *AER*, 1963.

10. The available unemployment data are hardly of any explanatory value, not only because of the problem of definition of unemployment in an agrarian African setting but also because most of the unemployed do not bother to register and there is no real incentive for them to do so.

never conflict; indeed, conflict is likely where foreign investors behave like opportunist speculators. However, a workable balance of interests, even if sometimes a delicate one, can be maintained between governments and foreign investors. It must, after all, be recognised that private capital seeks profit and tends to operate on commercial lines. As a one-time ICI chairman, Sir Paul Chambers, said, 'It is not part of the duty of any private enterprise company to use the funds of the stockholders to help the development of an underdeveloped country in such a way that the profits accruing to the shareholders are less than if the funds were used in some other way.'[11] Be that as it may, any meaningful solution to this problem must place greater emphasis on the development needs of the host country. Individually, many LDCs often find themselves in a relatively weak bargaining position *vis-à-vis* the ubiquitous transnational firms, when it comes to keeping the operations of the latter in check. But it is hoped that economic integration will improve the bargaining position of members when they deal not only with multinational companies but with all third parties.

Another factor militating against the expansion of the modern sector at a national level stems from the increasing competition between the countries of the sub-region. Projects exist for the establishment of new factories even when existing factories producing the same products in neighbouring countries are operating at less than full capacity. In some cases, such as Ghana from 1961 through 1965, under-utilisation of capacity was also due to import restrictions which reduced the supply of necessary imported raw materials and components, but in many cases it is a reflection of the individual countries' desire to expand production to meet national demand and, within groupings like the defunct UDEAO, to leave a comfortable surplus for export to members of the union.

There are, for example, a number of bicycle assembly factories in West Africa, located in Nigeria, Ivory Coast and Upper Volta. Although some of them operate at less than full capacity, new bicycle assembly plants are now springing up in other parts of the region. Existing breweries in the UDEAO countries operate at 50 per cent capacity, yet new brewery plants are going into production. Similar examples can be given for cigarettes, shoes, soap, matches and so on. The same problem has arisen with regard to intra-regional trade in agricultural products.

A case in point is to be found in animal products, the traditional

---

11. P. Adamson, *UNCTAD 3: Make or Break for Development*, Communications Ltd., 1972, 22. But whether this is an acceptable policy in the light of development problems of LDCs today is debatable.

export of the landlocked to the coastal countries. These are now being produced, through official protection and subsidy,[12] by the coastal countries at the expense of their suppliers in the savannah zone. Ironically enough, this has been happening in the name of diversification policy, a new article of faith for primary producers, which derives much of its force from production possibilities without due regard to the prospects for demand.

3. *Patterns of production and trade*

The patterns of production and trade have a very important bearing on the question of integration in West Africa. Despite the continued importance of the subsistence sector, the sub-region is highly dependent on foreign trade. Historically (see Chapter I), foreign trade has been the dominant influence on the development of West African economies. Their dynamic, monetised sector has grown out of and is still largely based on the production of a few tropical agricultural products and minerals for export to the industrialised countries, chiefly those of Western Europe (Table II.4).

On a continental scale, Professor Robson has sought to identify the origins of the African patterns of trade and development. He distinguishes two main patterns of development in African economies.[13] According to this classification the first group of economies developed mainly on the basis of a growth of commercial mining or European agriculture, stimulated by large-scale capital inflows from abroad. These economies, it is argued, were characterised by the existence of European settler communities and Africans participated in economic life mainly as wage earners. Algeria, Zimbabwe, Zambia, Kenya and Zaire are among the best-known examples in the category.

The second group of economies, like those of Ghana and Uganda, grew on the basis of development of a peasant agriculture, producing cash crops for export. It is further pointed out that the economies falling into the first group have become more industrialised and have enjoyed faster rates of growth than the latter; this is partly explained by the concentration of foreign investments during the colonial period in these areas. This would also mean that income would be less evenly distributed, and economic activities would tend to polarise around the major cities in the absence of countervailing

---

12. Tariff increases by Nigeria in 1966 on meat and meat products could be said to be 'clearly protective to encourage local production'. See Economist Intelligence Unit, *Quarterly Economic Review* (Ghana, Nigeria, Sierra Leone and Gambia), London, April 1966.
13. P. Robson, *op. cit.*, 67.

Table II.4
WEST AFRICA: DESTINATION OF MERCHANDISE EXPORTS
(% shares)

	Industrialised market economies		Sub-Saharan African countries		Other LDCs		Centrally planned economies		Capital-surplus oil exporters	
	1960	1979	1960	1979	1960	1979	1960	1979	1960	1979
Benin	90	89	8	2	0	8	2	1	0	0
Gambia	97	93	3	1	0	6	0	0	0	0
Ghana	88	70	2	2	3	15	7	13	0	0
Guinea	63	69	10	3	9	26	18	0	0	2
Guinea-Bissau	n.a.	29	32	22	n.a.	38	n.a.	1	n.a.	0
Ivory Coast	84	78	3	6	13	11	0	5	0	0
Liberia	100	86	0	0	0	14	0	0	0	0
Mali	93	68	7	15	0	17	0	0	0	0
Mauritania	89	88	11	2	0	9	0	0	0	1
Niger	74	97	26	1	0	0	1	0	0	2
Nigeria	95	87	1	2	3	11	0	0	0	0
Senegal	89	59	4	27	7	14	0	0	0	0
Sierra Leone	99	98	1	1	0	1	0	0	0	0
Togo	74	67	26	8	0	17	0	8	0	0
Upper Volta	4	75	96	9	0	16	0	0	0	0

Source: World Bank, *Accelerated Development in Sub-Saharan Africa: an Agenda for Action,* 154.

policy measures. But in the second case where the rates of growth are slower because the techniques of production are simpler, the income distribution seems more even.

Adopting this dichtotomy only as a first approximation,[14] we can say that West Africa falls within the second category. But even so, it is also clear from this classification that both groups share heavy dependence on foreign trade. And it is this export-oriented development together with the allied problem of low capacity-to-import, a structural but intractable phenomenon among African countries, which imposes one of the severest constraints on their rate of economic growth.

This high rate of economic dependence has been further aggravated by the limited range of primary exports in most of the countries — in contradistinction to the wide range of their imports of manufactures and foodstuffs. One or two commodities contribute the major part of export receipts. In eight of the fifteen countries, a single commodity group accounts for over 70 per cent of total exports and in the rest two commodity categories earn that much (Table II.5). In some extreme cases a single commodity earns the bulk of the country's foreign exchange. Consider that for Ghana, cocoa accounts for about 65 per cent of export income; Liberia, iron ore (60 per cent); Senegal, groundnuts (70 per cent); Nigeria, petroleum (90 per cent); and Mauritania, iron (84 per cent). Given the instability and unplanned character of the capitalist markets of Europe and North America, which are the major buyers of West African exports, and the competitive efforts of the developing nations to increase their output in the face of low income elasticity of demand for most of West Africa's exports and of high income elasticity of demand for manufactures, the long-term prospects in terms of existing patterns for the earning of foreign exchange, and hence development, are poor (Table II.6).

Aside from the foregoing, intra-sub-regional trade has one more outstanding feature. It is that West African countries are principally not 'each other's customer'. The share of their mutual trade in total

---

14. As Professor Robson himself emphasises, this dichotomy is useful only as a first approximation. Some African economies can conveniently come under any of the two categories. Ivory Coast is a case in point: it is one of the fastest growing economies in tropical Africa but mining has little or nothing to do with it. The country has few French settlers but they are as active in industry as they are in agriculture. Dakar (Senegal) was until independence the administrative centre for all French West Africa but it does not seem to be growing as fast as Abidjan, and certainly not faster. Furthermore, although West Africa as a whole does not have white plantation enclaves of the East African variety, there are inequalities in development both within and between the countries of the sub-region.

Table II.5
WEST AFRICA: DEPENDENCE ON MAJOR EXPORT CATEGORIES, 1978

Food & Beverages	%	Minerals & Metals	%	Fuels	%
Guinea-Bissau	96	Guinea	98	Nigeria	91
Gambia	90+	Mauritania	87		
Ivory Coast	74	Liberia	63		
Ghana	73	Togo	49		
Senegal	72	Niger	40		
Upper Volta	49				
Mali	47				
Sierra Leone	47				
Benin	40				

Source: World Bank, *op. cit.*, 152.

Table II.6
PROJECTED PRICE, VOLUME AND VALUE OF PRIMARY PRODUCTS OF EXPORT INTEREST TO WEST AFRICA, 1990

	Price index (1980 = 100)	Volume index (1980 = 100)	Value index (1980 = 100)
Minerals & Metals			
Iron Ore	115.8	133.9	155.1
Bauxite	105.6	199.5	210.7
Phosphate rock	117.4	164.9	193.6
Tin	95.0	110.6	105.1
Food & Beverages			
Coffee	96.7	125.6	121.5
Cocoa	66.2	143.5	95.0
Groundnuts	142.9	84.0	120.0
Groundnut oil	134.8	122.5	165.1
Palm oil	131.0	220.0	288.2
Bananas	90.6	135.6	122.9
Nonfood Primary			
Timber	135.7	138.3	187.7
Cotton	125.1	110.5	138.2
Rubber	109.4	140.3	153.5
Fuel			
Petroleum	137.0	123.6	169.3

Source: World Bank, *op. cit.*, 22.

external trade is negligible, a mere 3.6 per cent for the sub-region as a whole[15] (Table II.1). The factors that have shaped the pattern of intra-regional trade are mainly of two kinds: traditional, climatically induced specialisation in the production of foodstuffs and certain agricultural materials; and the existence of preference systems and monetary arrangements among groups of West African countries. Table II.7 suggests that a sizeable proportion of intra-zonal trade consists of foodstuffs. Even allowing for their relatively large volume of mutual trade in foodstuffs, the countries of the sub-region are able to supply to each other on the average only around 7 per cent of their total food imports. But despite the limited scale of their intra-zonal trade, it is nevertheless very important for the landlocked countries whose major exports include live animals and meat.

The only crude material of import entering intra-zonal trade is petroleum. With Nigeria's oil bonanza, petroleum has become an important item in trade. Nigeria now supplies a significant proportion of the crude oil entering West African trade (i.e. it sells crude petroleum to the refineries in Ivory Coast, Ghana, Senegal, Liberia and Sierra Leone) — and the ratio is likely to go up in the near future. As for sub-regional exchanges in manufactures they add up to only about 1 per cent of total imports of manufactured goods. This figure includes some re-exports of imported manufactures by the coastal countries to their landlocked neighbours.

The import structure of West African countries indicates that there is considerable potential for intra-regional trade in manufactures and, *ipso facto*, a considerable scope for import substitution industrialisation mainly in consumer manufactures and machinery and transport equipment (Table II.7). Even so, the demand for domestic manufactures is restricted by the small market sizes of most West African states. Evidently, in order to exploit the import substitution potential of the entire region, there must be systematic planning of optimal market sizes to ensure adequate demand either on a zonal basis or with the whole region as a single market for particular product items.

It follows from the foregoing that whatever manufacturing production is taking place in West Africa at the present time is essentially for domestic consumption within each country. Intra-regional trade in manufactures is negligible, and none exists between the region and the rest of the world. West African countries should

---

15. The intra-sub-regional trade figure is much lower than the continental average. Total intra-African exchanges for 1960–6 (average) stood at 7 per cent. The comparable figures for Latin America and Asia are 9 per cent and 25 per cent respectively.

Table II.7
WEST AFRICA: COMMODITY COMPOSITION OF MERCHANDISE IMPORTS, 1960 AND 1978

	Food		Fuels		Other primary commodities		Machinery and transport equipment		Other manufactures	
	1960	1978	1960	1978	1960	1978	1960	1978	1960	1978
Benin	17	15	10	15	1	2	18	22	54	46
Gambia	n.a.	24	n.a.	9	n.a.	3	n.a.	14	n.a.	50
Ghana	19	9	5	16	4	5	26	26	46	44
Cape Verde	n.a.	n.a.	n.a.	n.a.	n.a.	n.a.	n.a.	n.a.	n.a.	n.a.
Guinea	n.a.	43	n.a.	5	n.a.	1	n.a.	29	n.a.	22
Guinea-Bissau	n.a.	13	6	10	2	2	27	39	47	36
Ivory Coast	18	17	4	18	7	1	34	32	39	32
Liberia	16	19	5	14	4	2	18	30	53	35
Mali	20	n.a	3	n.a.	3	n.a.	39	n.a.	50	n.a.
Mauritania	5	10	5	12	4	2	18	33	49	43
Niger	24	14	5	2	6	2	24	44	51	38
Nigeria	14	23	5	12	2	21	19	18	44	26
Senegal	30	21	12	12	5	1	15	24	45	42
Sierra Leone	23	8	6	14	3	4	32	37	43	37
Togo	16	19	4	9	1	0	24	43	50	29
Upper Volta	21									

Source: World Bank, *op. cit.*, 151.

strive not only to achieve import-substitution which, as Little, Scitovsky and Scott[16] demonstrate, has its limits, but they should also aim to expand exports in manufactures to other countries outside the region.

Given the small size of most West African countries and their concomitant open nature and heavy dependence on export-import trade (see Table II.8), there is a compelling need to insulate these economies from the dangers of instabilities traceable to external disturbances. It is therefore expedient to hasten the move towards economic integration to reduce the vulnerability of individual members of the region.

Table II.8
OPENNESS OF THE ECONOMIES OF WEST AFRICA, 1979

	Exports(X)/GNP	Imports(M)/GNP	(X + M)/GNP
	(1)	(2)	(3)
Benin	22.4	42.0	64.4
Cape Verde	n.a.	n.a.	n.a.
Gambia	60	50	110.0
Ghana	24.2	22.0	46.2
Guinea	25.1	23.4	48.5
Guinea-Bissau	10.3	44.9	55.2
Ivory Coast	29.5	29.2	58.7
Liberia	56.2	54.1	110.3
Mali	18.6	18.9	37.5
Mauritania	28.7	50.6	79.3
Niger	8.3	13.3	21.6
Nigeria	32.7	22.4	55.1
Senegal	17.8	32.0	49.8
Sierra Leone	24.1	34.9	59.0
Togo	29.9	52.5	82.4
Upper Volta	8.0	25.2	33.2

Note: Column 3 denotes the *degree of dependence* on trade (openness) which is generally high for small countries and/or countries with special external trade links.
Source: Computations based on World Bank, *Accelerated Development in Sub-Saharan Africa: Agenda for Action*, 1981, 143, 149.

The other factor affecting the geographical pattern of trade in West Africa derives from the existence of a preferential system and of monetary arrangements among most of the Francophone countries of West Africa. Reasons for this are essentially historical,

---

16. I. Little, T. Scitovsky and M. Scott, *Industry and Trade in Some Developing Countries: a Comparative Study*, Oxford University Press, 1970, 59.

as noted earlier. Exchanges among members of the defunct West Africa Customs Union (UMAO), which will be discussed later, greatly exceed trade between that group and the rest of the sub-region. The exports of other West African countries to the UDEAO group have to clear the latter's non-preferential external tariff of 5-25 per cent to which fiscal levies and supplementary taxes are added. Perhaps, no less important than tariff preferences in stimulating trade among UDEAO members and in discouraging the latter's trade with the rest of the sub-region are the existing monetary arrangements. The UDEAO members have a common currency, the CFA franc (see p. 38, below), which is automatically convertible with the French franc,[17] whilst the separate currencies of several other West African countries are not freely convertible into CFA francs or into each other. As if this were not enough, some of the latter countries maintain quantitative trade restrictions.

However, with the formation of ECOWAS and the signing of the Lomé Convention there are now forces at work which may reduce or possibly eliminate these barriers to increased intra-zonal trade in the not too distant future.

4. Conditions for effective integration in West Africa

On a rather general level, members of a prospective economic grouping would require satisfactory assurances on three broad fronts prior to full membership. First, each participating country should believe that the benefit accruing to it — regardless of how this was calculated for the short term and the long term — will be greater than anything that could be achieved by remaining outside the scheme. Secondly, the members of the grouping should be, individually and collectively, very willing to make the necessary sacrifices and compromises towards the adopting of certain policies to realise the aims of the community, and finally the physical infrastructure, especially transport, should be made efficient enough to facilitate a well-ordered intra-zonal distributive network. The issues involved in these pre-conditions are interrelated and their exposition somewhat twisted. Political considerations enter even more strongly into these matters. In what follows we shall reorganise them and try to highlight the more important questions affecting effective integration in West Africa.

The first factor relates to similarity in the levels of development among the prospective members. It is often argued that the extent of the benefits which an integration scheme can bring depends largely on the economic development already achieved by the partner

17. IMF, *Surveys of African Economies*, vol. 3, Washington DC, 1970, 24.

countries, on the form of their development, particularly industrial development.[18]

Other factors — most of which have been discussed above, e.g. size of their subsistence sector, their natural resources, climatic conditions, and the supply of labour and capital — are also important. But here we want to concentrate on the place of the size of the industrial sector in integration schemes. It goes without saying that where one or more of the integration partner countries are relatively large in the industrial sector, the partner countries may well resist the introduction of free trade and equal competitive conditions. They may, with good reason, fear that their own industries would not, if unprotected, be competitive in the area or that, given the tendency of industries within an integrated area to cluster in a few industrial growth points, their nascent industrial sectors would be unable to expand.[19] Thus, *a priori*, it seems that integration is most likely to be successful where industrial sectors are of a similar size and composition.

In West Africa, as we have noted in the previous sections, the size of the industrial sector (excluding extractive activities) is very small both in relative and absolute terms (Table II.3). Although some countries (Ghana, Ivory Coast, Nigeria and Senegal) have rapidly expanding industrial sectors, yet their manufacturing output is scarcely up to 20 per cent of GDP. This, of course, is not to say that the level of development is everywhere the same but that the existing differences in structure and development seem to be within manageable proportions. Some kind of industrial rationalisation will certainly be needed in the less efficient countries before they can allow free trade in the products of their threatened industries. Equally important here are the differences in tariff structure existing between prospective members before integration. We will discuss this in detail later in the study but it is necessary to underline the fact at this juncture that the larger the differences in tariff structures before union, the more difficult it will be to agree on the reduction and eventually the removal of tariff protection among members.

West African countries, it would seem, stand a fairly good chance

---

18. Compare with F. Andic *et al.*, *A Theory of Economic Integration for Developing Countries*, London: George Allen and Unwin, 1971, 45.

19. This form of resistance has been very much in evidence in LAFTA. The failure of LAFTA to reach agreement in 1967 on the second stage of drawing up the 'Common List' of products to be unconditionally freed of trade restrictions at the end of the transition period in 1973 was a glaring case. The Arab Common Market, where Egypt has a more highly developed industrial sector than any other, has had the same trouble.

of meaningful integration, if only because of the small size of their industrial sectors. For one thing, the small industrial sector means small vested interests that might oppose integration for fear of the painful adjustments possibly resulting from the removal of protection inside the integration area. However, if, as often happens, these vested interest are backed by governments, local or foreign, they might succeed in imposing exceptions for themselves that would reduce the scope of possible benefits and the speed with which they can be obtained. Secondly, the smaller the industrial sectors at the inception of integration, the larger the scope for capturing the benefits of specialisation through regional investment planning.[20]

But it must be noted that groupings in which one or more countries have a significantly larger and more efficient sector than the others, like Ivory Coast in the Entente States, can still bring benefits. Provided the larger countries do not produce all the goods which could be economically produced for the integrated market, a regional policy for investment planning and a satisfactory compensation arrangement could persuade the less industrially developed partner-countries to abolish their trade barriers.

Another problem which West Africa faces in relation to integration arises out of the obstructing effects of extra-African politico-economic ties and cleavages. On the economic front the economies of Francophone West Africa, except Guinea, are still closely meshed with those of the members of the European Economic Community (EEC), in particular France. The fact that all West African countries are now associated with the EEC hardly alters the picture. This is essentially the by-product of colonial history but those historical relationships have been kept alive till now by the post-independence association with the EEC, first through Yaoundé, and now with the Lomé Convention.

Whatever form economic co-operation in West Africa or part of it may eventually take, promotion of trade among the countries of the subregion is clearly one of the major objectives. This is borne out by the provisions of the Treaty of the Economic Community of West African States (ECOWAS). In brief, the Treaty provides, among other things, for the establishment of customs union among the

---

20. The task of co-ordinating regional investment planning is by no means an easy job especially in a region where competing industrial sectors work below full capacity in the partner countries. The Maghreb countries are known to have faced this difficulty. (See F. Kahnert et al., *Economic Integration Among Developing Countries*, Paris: OECD, 1969). On a smaller scale a similar situation is developing in West Africa).

member-states through the progressive elimination of tariff and non-tariff barriers to trade (Article 12).[21]

Fortunately the current Lomé II Convention between the EEC and the African, Caribbean and Pacific Group of States (ACPs), which supersedes earlier agreements (viz. the Lomé I and Yaoundé Conventions), accorded priority to the encouragement of regional groupings among LDCs, especially ACP members. The importance of providing a special Fund for the promotion of regional co-operation was recognised long before the signing of the Lomé I Convention. On the eve of Lomé I, the ACPs requested a special regional development fund which the EEC turned down, but in the end 10 per cent of European Development Fund IV, or 300 million units of account (in two instalments), was set aside for regional and inter-regional projects between the ACPs and between the ACP and non-ACP LDCs.[22] Where a non-ACP state, either on its own or via a regional organisation, becomes a party to either type of venture, the Convention requires it to organise its own share of the project cost. This is because the European Development Fund (EDF) is strictly confined to arranging the contributions of the ACP states, for which it hardly provides enough.

Although Lomé II has doubled the size of available aid for such projects (i.e. 600 million units of account),[23] it is unlikely to prove sufficient. By the end of 1977 the EDF had received project requests totalling 940 million units of account when no more than 300 million units of account were actually provided under Lomé I (for the period 1975-9). Thus, it can be argued that the demand of the ACPs at the negotiations on Lomé II for 1,000 million units of account was by no means excessive, and represented the minimum amount required to make some headway in fostering inter-regional co-operation. Regrettably, the ACPs got only 600 million units of account for the period from March 1980 till March 1985. This amount will assuredly not be sufficient to exploit the opportunities for the mobilisation of considerable co-finance to promote the large-scale projects which are at present necessary if regional and inter-regional co-operation are to be advanced.

Apart from the actual amount of aid provided, it is concentrated on regional and inter-regional ventures in which only ACP states are participants; and it has not been fairly distributed. As the ACP

---

21. For details see Chapter 6 of this book. Also see Federal Republic of Nigeria, *Third National Development Plan, 1975–80*, vol. 1, Lagos, March 1975, 36.

22. European Economic Community, *The Lomé Convention and Related Documents* (Brussels, 1975).

23. *The Second ACP-EEC Convention and Related Documents* (Brussels, 1980).

Council of Ministers noted at its Lusaka meeting in December 1977, Central, Eastern and Southern Africa had received only 10 per cent of the allocation from the first instalment under Lomé I.[24] An effective political formula for fair distribution of available funds, without compromising development and economic criteria in aid allocation, will enhance the value of such assistance. Perhaps the Joint ACP-EEC Committee set up to review the Financial and Technical Assistance Title of Lomé will come up with a satisfactory solution to the problem.

In the political arena, the Lomé Convention offers the EEC countries considerable scope for political and diplomatic pressures on the ACPs. This consideration, however, falls outside the scope of this study; suffice it to say that the former colonial powers in Europe (especially France), which have influence on West African countries, could apply their influence — when their interests are involved — in such a way as to impinge on the freedom of the West African countries in deciding whether to join or remain in a given regional scheme. Indeed, it would appear that in spite of the OAU, the Lomé Convention and ECOWAS, Francophone West Africa still wants to maintain its *affinités complémentaires* with France, even if sometimes at the expense of West African co-operation interests.[25] Even so, everything depends on everything else. Once a decisive community spirit emerges at the regional level, the effects of external influences and pressures will be considerably reduced.

A third prerequisite for effective economic co-operation in West Africa turns on transport and communications. By definition, economic integration between any group of countries implies easy access to each other's markets. High intra-regional transport costs may give 'natural protection' to a number of small-sized plants which outweighs the benefits of economies of scale; by the same token, poor communications, involving time-consuming procedures and insufficient information sharing, can only retard economic growth.

As indicated earlier, the traditional transport structure of the sub-region is based on the need to move relatively bulky primary materials to a major port. Thus, with the exception of landlocked countries, road and rail links normally lead from the interior to the coast while shipping routes connect ports to developed countries' markets. It has long been recognised that 'an integrated transport system' is the key to a rapid and comprehensive expansion of intra-

---

24. European Economic Community, *Second Annual Report of the ACP-EEC Council of Ministers* (Brussels, 1978), 64–5.
25. See *West Africa*, 5 May 1975, 505.

West African trade and industry and that the main feature of the transport system is 'the absence of satisfactory links between countries and territories'.[26] Improved transportation aids trade in two ways: first, it allows for expansion of trade in products currently produced in the sub-region, and secondly, it encourages the growth in trade of new goods from subregional industries the establishment of which will become possible as a result of integration.

Efforts geared towards the co-ordination of a new subregional transport network have been noticeable in the past. Several studies have been made, but little or no performance has followed. At the first session of the West African Transport Conference in Monrovia in 1961, agreement was reached, albeit tentatively, on a network of subregional roads. With later supplementary proposals by the ECA,[27] this network provides for one road link between each neighbouring country and covers a total length of approximately 20,000 km. (12,500 miles). An effort was made in 1964 by the ECA and the International Civil Aviation Organisation (ICAO) to promote the co-operative development of air transport. Countries agreed on the need for co-operation, but no further action has been taken. An estimate was made of the modifications to the existing West African transport system which would be needed if the recommendations of the 1963 West African Industrial Co-ordination Mission were implemented.[28]

It is of course true that new transport flows in the sub-region cannot be determined with any precision until actual decisions have been made on the siting of industries. However, the concentration of the market around the Bight of Benin means that this area is likely to be the first choice of large-scale industries, and it was suggested that studies be made concentrating on the improvement of the links between these countries and the rest of the sub-region. It was therefore recommended, first, that road networks be improved, mainly between Nigeria and Niger, Ivory Coast and Upper Volta and Mali, and Ghana and Upper Volta; secondly, that a Ghana-Togo-Benin-Nigeria rail link be developed; and thirdly that railroad extensions from Maiduguri in North-Eastern Nigeria to Fort Lamy (Chad) and from Ivory Coast into Mali be considered after a few years.

On the basis of an ECA report and three other studies,[29] some

---

26. ECA, *Report* on the First Session of the West African Transport Conference (E/CN. 14/147, 1961), 4.
27. ECA, *Inland transport in West African subregion* (E/CN. 14/TRANS/17, June 1964).
28. ECA, *op. cit.* (ST/ECA/109), 69.
29. These include two bilateral studies of transport development — one by a team from the West Germany: 'Transit problems of African land-locked states'

conclusions could be drawn on the perspectives of a West African transport system. First, it is highly probable that most of the large-scale industry in any subregional plan will be located along the coast, and the expansion of coastal shipping on a co-operative basis should help integration in the area. Secondly, the expansion of inland transportation would more profitably concentrate on road facilities except for a few special cases such as a coastal rail link between the Ivory Coast and Nigeria. Thirdly, it appears that most of the investment in the road network should be in the improvement and proper maintenance of existing routes with particular attention being given to the standardisation of transport regulations and administration.

More recently, some more concrete steps have been taken towards transport development in West Africa. Following the formal ratification of the ECOWAS Treaty in 1975, a conference of West African Transport Ministers was held in Abidjan, Ivory Coast, at the request of the ECA and OAU to lay the foundations for a co-ordinated regional transport system. Decisions taken at this conference covered the development and the financing of an integrated system of roads, railways, maritime and coastal shipping, inland waterways and air transport in the sub-region and the standardisation of transport regulations and administration.[30]

In addition to the intra-West African transport development plans, the continental transport projects would benefit the region too. The Trans-Saharan road (otherwise known as Unity Highway) that will meander through the Algerian desert via Tamanrasset in the southern tip of that country, is planned to terminate in Kano. Also the Trans-African Highway, which will link Mombasa on the Kenya coast in the east with the western parts of the continent, will have Lagos as its terminus; and Nigeria is earmarked to be the nerve centre of the rapidly growing coastal networks that will eventually serve all the countries that form ECOWAS.[31] Although geophysical obstacles have often made transport connections prohibitively expensive, the logic of economic integration in the sub-region demands a balanced transport network; and the challenge, as the foregoing shows, is being met.

The fourth problem likely to affect the expansion of intra-West African trade revolves around the absence of an all-embracing region-wide payments arrangement or union. Before the countries

(E/CN.14/TRANS/28, Aug. 1965); one by a French team: 'Aspects of transport development in West Africa' (E/CN.14/INR/118/ADD.1, Oct. 1966), and an EEC study of December 1966, 'Possibilités d'industrialisation des Etats. Africains et Malgache associés'.

30. See *West Africa*, 21 July 1975.
31. *Africa Magazine*, October 1978.

of West Africa became independent, there were two main financial zones: the French franc for the French-speaking countries and the British pound sterling for the English-speaking countries.

For the Francophone countries the colonial currency unit in use was the CFA (*Colonies Françaises d'Afrique*) franc. The CFA franc, which had parity with the French franc, was issued by each country and was legal tender only within its country of issue. In 1962 a new monetary arrangement, the West African Monetary Union, which will be further discussed elsewhere, came into existence. By this arrangement Benin, Ivory Coast, Mauritania, Niger, Senegal, Togo and Upper Volta use the same currency issued by a central bank, the Banque Centrale des Etats de L'Afrique de L'Ouest (BCEAO). The new CFA franc notes (these initials now signify Communauté Financière Africaine) issued by the BCEAO are legal tender in all the member-countries, unlike the colonial CFA franc. Also the BCEAO notes have a fixed parity with the French franc, and their convertibility is guaranteed by the French government hence membership of the franc zone affords the countries involved a high degree of monetary stability and a relatively high level of intra-zonal trade. But they cannot pursue independent monetary and financial policies. Such policies are controlled and directed by France.

The English-speaking countries of West Africa — the Gambia, Sierra Leone, Ghana and Nigeria belonged to the pound sterling area. These countries were monetarily dependent on Britain as the Francophone states were on France, but were loosely attached to the pound sterling. For, unlike the franc zone, the sterling area countries pursued independent foreign exchange and reserve holding policies, diversifying their reserves in foreign currencies of their choice. The monetary union embracing the Anglophone states of West Africa operated under the West African Currency Board. The Board provided a simple but effective method of managing a separate colonial currency and of maintaining complete stability in the rate of exchange between the colonial currency unit, which it issued, and sterling. Although the West African Currency Board was dismantled during the course of independence of the respective members and the establishment of national monetary institutions, the member-countries still belonged to the sterling area until the latter was terminated by the United Kingdom Government on 23 June 1972, which thereby reintroduced intra-area payments difficulties.

Article 37 of the ECOWAS Treaty provides for the establishment 'in the short term, of bilateral systems for the settlement of accounts between the member states and, in the long term, of a multilateral system for the settlement of such accounts'. Fortunately, progress in this area has been faster than was envisaged by the Treaty. The idea

Conditions for effective integration in West Africa 39

of a multi-lateral payments system in the form of an Association of African Central Banks has been toyed with since 1963, 'when African Heads of States and Governments, meeting in their first summit conference in Addis Ababa, unanimously resolved to set up a preparatory economic committee to study, in collaboration with the governments, and in consultation with the Economic Commission for Africa, a large range of monetary and financial questions.'[32] Subsequent efforts in this direction culminated in the establishment of the West African Clearing House which came into legal existence on 25 June 1975 and started operations on 1 July 1976.

The objectives[33] of the West African Clearing House are:
— to promote the use of the currencies of the members of the Clearing House for sub-regional trade and other transactions;
— to bring about economies in the use of foreign reserves of the members of the Clearing House;
— to encourage the members of the Clearing House to liberalise trade among their respective issues; and
— to promote monetary co-operation and consultation among members of the Clearing House.

A number of bilateral payment arrangements had been in existence for quite a long time in the region. Some countries resorted to bilateral trade and/or payments agreements in order to minimise the difficulties stemming from rigid exchange controls which are common in the area. Such bilateral arrangements exist between Ghana and Mali, Upper Volta and Togo, Niger and Benin, Nigeria and Niger, Togo and Mali, and between the Gambia and Senegal. There is also a bilateral clearing between the Central Banks of Ghana, Nigeria and Sierra Leone for embassy transactions.

With the establishment of the West African Clearing House, it is hoped that all bilateral arrangements would terminate as soon as they expire, thereby giving way to a smooth flow of trade and payments throughout the whole region. This, however, is still a hope. For one thing, although membership is open to all West African countries, it remains optional. Mauritania and Guinea are still to join the clearing arrangement, although their membership of ECOWAS will oblige them to do so. Besides, the coverage of transactions handled by the clearing house was considered within the framework of the peculiarity of trade in the sub-region, namely limited volume of intra-area trade; narrow composition of trade

32. The West African Clearing House, *An Experiment in Multilateral Co-operation in West Africa*, Freetown, 1976.
33. *Ibid.*, 4.

items, especially reliance on one or two commodities by some countries, widespread unofficial frontier trade transactions, which are not easily amenable to official control; and goods that are not of the region's origin.

The net effect of these structural problems is to limit the effectiveness of the West African Clearing House as a mechanism for rapid intra-West African trade. For example, the period for settlement of claims is fixed at one month, and in order to reduce the frequent recourse by central banks to settlements in foreign exchange, each participant is required to extend to the rest of the sub-region as a group a monthly interim credit line equivalent to only 20 per cent of its annual intra-regional trade.[34]

Nevertheless, the advantages of clearing and payments union are many for intra-union trade among LDCs.[35] A simple payments union provides automatic credit to finance all or part of a member's deficit with the other members of the union, thereby multilateralizing credits among its members. Also, in so far as member-countries do not spend their foreign exchange in intra-zonal transactions, it would permit some of them to hold back their reserves for trade with third countries. In other words members experiencing deficits with their partners in the union can finance their deficits without drawing upon their reserves. Furthermore, if members are not obliged to earmark their reserves for settlements within the union, they would have less incentive to restrict their imports from extra-union sources, hence, the expansion of intra-union trade need not necessarily mean a reduction in the volume of imports from the outside world.

However, the argument has its reverse side. A member which has a surplus with the union could not use its surplus to settle a deficit with the rest of the world. On the other hand, if it had a deficit with the union, it would be obliged to curtail its imports from outside or to draw on its reserves. Given the low level of intra-African trade, it is argued that the feedback effect of a payments union might be detrimental to a LDC which depends largely on the imports of capital goods from outside for its industrialisation programme. While the basic fact here is clear, the desirability of some kind of payments

---

34. See S.B. Falegan, 'Clearing and Payments Arrangement within the West African sub-region', paper presented at the ECOWAS conference, Lagos, 23–27 August 1976.

35. For further reading see R.F. Kahn *et al.*, 'The Contribution of Payments Arrangements to Trade Expansion' in P. Robson (ed.), *International Economic Integration*, Harmondsworth: Penguin Books, 1971. See also P. Robson, *op. cit.*, 287–91; UN Committee (1966), 'Trade Expansion and Economic Co-operation among Developing Countries', and UN Committee (1965), 'International Monetary Issues and the Developing Countries'.

arrangement for the promotion of intra-regional trade cannot be in dispute — provided the parties to any such arrangement are determined to see it work and would be willing to make the necessary sacrifices. For example, members can set up 'a common pool of foreign exchange' made up of contributions on a quota basis, for the purpose of enabling deficit countries to make intra-union payments in foreign currency over and above an agreed maximum figure. Figures up to or below the maximum would be settled in local currency. This source of credit, which could be administered either by a co-operation of members' central banks or by a special union bank, certainly holds out some hope for the expansion of intra-zonal trade in West Africa.[36]

Such other factors as compensation strategy, which directly affect the success of integration schemes, will be treated separately later in the study.

In conclusion we should mention here some non-economic centrifugal forces operating on the sociocultural sphere, which can undermine co-ordinated economic relationships. The African languages spoken in West Africa are as many as 150.[37] Religions too are numerous, and nationalism (or what Western writers refer to as tribalism) is strong in the larger communities. Customs differ widely within each country and among countries, while the level of literacy varies in the same vein. Political systems pretend to Western-oriented liberal democracy, but they are still 'unpolished' and undergoing an evolutionary process. Needless to say, the political stability of some of the regimes in the sub-region, especially the military juntas, is very uncertain. English or French is spoken by the literate West Africans, but inter-personal and inter-country contacts, despite some marked improvements since the formation of ECOWAS, are still hampered by the Anglo-French cultural divide bequeathed by the colonial system. The strains and stresses arising out of these non-economic factors, while not insurmountable, have tended to reinforce the economic obstacles to integration.

---

36. This type of scheme would be without prejudice to the members' membership of the IMF and the credit facilities available to them as a result. See P. Robson, *op. cit.*, p. 291. But it is important to underline the point that the credit facilities available to the LDCs from this body *vis-à-vis* their needs are highly limited, not to mention the bureaucratic procedure and terms of granting these facilities. See T. Hayter, *Aid as Imperialism*, Harmondsworth: Penguin Books, 1971, chapter 2.

37. See Daryll Forde (ed.), *Ethnographical Survey of West Africa*, London: International Africa Institute, vols I-XIII.

III
THE THEORY OF INTEGRATION AND AFRICAN ECONOMIES

In many parts of the world to-day some form of economic integration is either in existence or actively in prospect. This widespread enthusiasm for the formation of customs unions, free trade areas, common markets or economic unions among groups of countries derives from a complex of motives and sentiments. In Western Europe after the Second World War, the search for a permanent peace and world order caused a spurt of integration initiatives some of which have been a success up to the present. For the Eastern bloc, socialist integration within the framework of the Council for Mutual Economic Assistance (CMEA) sets itself the aim of achieving rapid economic, scientific and technical progress in all the member-countries and of raising the material and cultural standards of their peoples through deep structural changes in the economy in accordance with the objective requirements of scientific and technical progress.[1] In LDCs also, political and psychological considerations have always been mixed up with what may be regarded as an alternative economic policy, designed in the hope of achieving satisfactory and lasting solutions to their trade problems.

The urge for close economic ties is further reinforced by the growing realisation, as demonstrated earlier, that the small size of some LDCs is a serious obstacle to rapid economic development.

But although, *a priori*, it is generally agreed that trade liberalisation within a grouping maximises economic efficiency from the group's point of view by comparison with a non-trade situation (autarky), it is possible and even likely that some countries or groups within each member-country will be hurt by the dismantling of trade barriers — despite the fact that the pie is larger for the members of the group as a whole. In other words, the gains from economic integration are usually achieved at the cost of distributive effects within the group. Thus the cardinal issue, to which we shall direct our attention here, is the extent to which the conventional theory helps to explain the effects of economic integration on actual or prospective members — the costs and benefits they derive from it,

---

1. CMEA, *Comprehensive Programme for Further Extension and Improvement of Co-operation and the Development of Socialist Economic Integration by the CMEA Member Countries* (Moscow, 1971), 9.

particularly in the case of trade groupings among LDCs.

The first part of this chapter will analyse the relevance of the conventional theory of integration to the countries of West Africa, in common with those of other groupings of LDCs, while in the second part we look at the role of measures of policy harmonisation within the framework of economic integration.

1. *The theory of integration and West African economies*

The economic analysis of the traditional theory of integration centres fundamentally on two things: (i) the effects of economic union on aspects of welfare; and (ii) the effects of union on the pattern and volume of trade. The former emphasises the welfare gains or losses from a marginal re-allocation of production and consumption patterns, under conditions of static equilibria in which such things as factor endowments, technology, demand, and population are assumed to remain unchanged. The latter, which focuses on the more dynamic aspects of the theory, such as the existence of economies of scale and variable factor proportions, relaxes the neo-classical static assumptions.

Taking its criteria for evaluation from the static effects,[2] the primary concern on the traditional theory of integration has been to examine the desirability of a customs union from the world's welfare standpoint. Viner's celebrated pioneer work[3] distinguishes between two effects: trade creation and trade diversion. Trade creation is said to occur if and when a pre-union high-cost domestic producer is displaced and replaced by an intra-union low-cost producer after the formation of a customs union. Since the formation of a customs union has brought in its wake the relocation of production in the lower-cost location within the union, trade creation has a salutary effect on the national income of the integrated economies and, *ipso facto*, of the world. Conversely, trade diversion occurs if, before the union, a high-cost (inefficient) producer having been sheltered by a post-union discriminatory tariff wall captures part of or the union market. This, in the traditional theorists' parlance, would represent not only a loss in world national income but a 'disaster' for specialisation on a world scale, and it may also be a loss or a gain to the total national income of the union. Thus in its simplest form the Vinerian analysis leads to one important conclusion, namely that a

---

2. Yu-Min Chou, 'Economic Integration in Less Developed Countries: the Case of Small Countries,' *Journal of Development Studies*, July 1967, 19.

3. Jacob Viner, *The Customs Union Issue*, New York: Carnegie Endowment for International Peace, 1950.

customs union raises the world's welfare if its trade creation effect outweighs its trade diversion effect. In other terms, trade creation, from the point of view of free trade, is a move in the right direction whilst trade diversion is a move in the wrong direction. The former occurs when the constituent economies are competitive in products, rather than complementary, prior to the formation of a customs union.

In many respects, the emphasis on the static effects in the traditional theory of economic integration is understandable. First, it is consistent with the static approach in international trade theory whereby economic integration is conceived as 'that branch of tariff theory which deals with the effects of geographically discriminatory changes in trade barriers'.[4] Secondly, and more important, it is consistent with the relative importance of adjustments likely to occur once a group of developed countries decides to integrate. Furthermore, the theoretical literature of economic integration dealt almost exclusively with customs unions of industrial economies, where the problem is not primarily one of economic development but one of relatively marginal adjustments in production and consumption patterns.

Before the publication of Viner's path-finding work,[5] the generally accepted view was that a customs union, since it represents a move towards free trade in the Haberler-Marshall sense, increases world welfare. It was in order to demonstrate the flaws implicit in this method of reasoning that Viner introduced the concepts of trade creation and trade diversion. Even so, although Viner's contribution has remained one of the important pillars on which customs union theory rests, and while we may derive some guidance from it on the effects of integration on production both outside and inside the integration area, it does not make it possible, as we shall see later, to judge the overall desirability of an integration scheme, especially in a developing region like West Africa.

Since the 1960s increasing attention has been directed to problems of economic integration among LDCs. Most writers[6] who consider

---

4. R.G. Lipsey, 'The Theory of Customs Union: a General Survey', *Economic Journal*, September 1960, 261-2.

5. J. Viner, *op. cit.*

6. See D. Seers, 'The Limitations of the Special Case', *Bulletin of the Oxford Institute of Economics and Statistics*, May 1963, 83; B. Balassa, *Economic Development and Integration*, Mexico, 1965, 16; T.A. Jaber, a Review Article, 'The Relevance of Traditional Integration Theory to Less Developed Countries', *Journal of Common Market Studies*, March 1971, IX, 3. On a broader plane the applicability of 'Conventional Economics' to LDCs has been called into question (see H. Myint, 'Economic Theory and the Underdeveloped Countries', *Journal of Political Economy*, 75, (1965).

these problems feel that the traditional theory of economic integration has limited relevance, if any, to LDCs. The several arguments which they use can be classified as follows. First, economic integration in the case of LDCs should be treated as an approach to economic development rather than as a tariff issue. Accordingly, it combines various aspects which could improve the international trade position as well as raise the level of economic development of LDCs. Secondly, the emphasis should be placed on dynamic rather than static effects in evaluating the desirability of economic integration among LDCs. The dynamic effects refer to the various possible ways in which integration affects the rate of growth of GNP of participating countries. They include (*a*) the economies of scale brought about by the enlargement of the size of the market for firms producing below optimum capacity before integration; (*b*) the external economies which shift specific or general curves downward; (*c*) the polarisation effect which refers to the cumulative improvement of the relative, or absolute, economic position of a member country or of some regions in the integrated area due to concentrated trade creation or attractiveness of labour and capital; (*d*) the effect on the volume and location of investment; and (*e*) the effect on economic efficiency and smoothness of trade transactions due to change in the degree of competition and change in the uncertainty and unilaterality of trade policies of individual countries.

The present economic structure, so the argument goes, is not acceptable and each LDC is trying to introduce positive changes. These changes are not marginal but structural. Their net effect will not be felt over a short period of time. Therefore, any evaluation of economic integration schemes should concentrate on the above potential or dynamic effects.

1(*a*). *Reformulation and extension of the theory*. In view of the structural characteristics of LDCs and to render the conventional theory of integration useful in judging the desirability of integration in LDCs, a reformulation and extension of the theory has been made. From the writings of the authors[7] who contributed to the reshaping of the standard theory for LDCs, three strands of thought can be distinguished. These are (*a*) that customs union theory should contribute to a more equitable distribution of income; (*b*) that trade diversion might be inevitable in a developing country; and (*c*) that the standard theory must incorporate dynamic aspects.

It is worth demonstrating at this point why these conclusions are necessary. First, Viner considers trade diversion a negative produc-

---

7. *Ibid.*

tion effect which necessarily reduces welfare. But it can be noticed that trade diversion is, at a basic level, taking place in individual countries level through import-substituting industrialisation policy which has become an article of faith in West Africa over the last three decades. The choice therefore is between trade diversion in favour of the most efficient producer in the region. In general, however, a pattern of industrialisation based on greater specialisation within the region will be more rewarding than one based on production by each country for its own domestic market, particularly for small countries.

Indeed, the key economic case for economic integration in West Africa rests on the potential for the exploitation of internal and external economies of scale, especially in manufacturing activities. Integration will affect the rate of growth of GNP of the participating countries, partly in the form of a more efficient scale of operation by existing enterprises, given the current level of under-utilisation of industrial capacity in the region, but primarily from a greater rate of investment in new industries. Besides, there are possibly benefits to be gained from specialisation between countries within the common market according to comparative advantage.

In West Africa, Nigeria and, to a less extent, Ghana can justifiably set up any heavy industry based on the home market.[8] Even so, this does not offend against the spirit and rationale behind the formation of ECOWAS. Even if the national market of any member-state of an integration scheme is large enough to satisfy the technical and economic optima of any kind of plant, there could still be gains from the enlargement of the market. In a larger market, like ECOWAS, more than one producer might be able to achieve the technical economies of scale, so that there could be additional economies arising from competition between them. Conceivably, a market of this size is not a sufficient condition for the benefits of competition between technically optimum-sized plants to be achieved; but to obtain the potential benefits of competition, a market which is a multiple of that required for the technical economies is necessary.

Secondly, in most LDCs there exists a situation of general low productivity, and in some sectors marginal productivity might approach zero. Also, as we noted earlier, unemployment is not uncommon. If trade diversion moves labour from low-productivity to more productive activities, it will bring about a gain in welfare. In LDCs, with considerable levels of unemployment as in West Africa this gain in welfare becomes more likely. The evaluation of integra-

---

8. See Uka Ezenwe, 'The Rationale of Economic Integration in West Africa', *Intereconomics* (Hamburg), No. 4, 1975, p. 107.

tion among LDCs should not therefore be confined to production and consumption effects; income and employment effects are equally important.

Thirdly, when the imports of LDCs are disaggregated, trade diversion appears to occur only in non-durable, and to a less extent in durable, manufactured consumer goods. In a static situation, no trade diversion or creation is likely to occur in their imports of capital goods which in most cases account for about 40 per cent of total imports. In a dynamic situation, it is argued that a higher rate of growth conceived by an economic integration scheme would require a larger investment. Since a large portion of this investment is imported as capital goods, the level of imports of integrated LDCs might then increase. In any case, the long-run impact of a regional trading arrangement is not to decrease trade with the rest of the world but rather to change its pattern and possibly to enlarge it.

Put another way, some writers argue that economic integration among LDCs should aim at trade diversion from DCs.[9] Hence the effectiveness of such economic integration then is to be indicated by the success of the trade diversion process. Professor R.S. Bhambri expresses this point in stronger terms:

It is . . . reasonable to suggest that trade diversion will be doubly beneficial. Firstly, by enlarging the size of market for manufactures in both countries, increased trade will help to reduce costs in industries where scale economies are important. Secondly, import substitution over a wider area will enable the region as a whole to spend a higher proportion of its foreign exchange on imports of capital goods and raw materials and help to increase the rate of investment and economic growth.[10]

Finally, trade creation, like trade diversion, should be looked at in dynamic terms. The dynamic trade-creating effect results from the increase in income of the integrated area and through the foreign trade multiplier. It is argued that this effect would be large enough to outweigh the dynamic trade diversion effect of economic integration among LDCs. As Kitamura aptly stated: 'The income effects, so far as trade with the outside world is concerned, will clearly tend to increase considerably the scope for beneficial exchange of goods with third countries, and this secondary trade expansion may very well more than offset the possible initial reduction of this particular type of trade.'[11]

---

9. On this, see S.B. Linder, *Trade and Trade Policy for Development*, New York: Praeger, 1967 and R.S. Bhambri, 'Customs Unions and Underdeveloped Countries', *Economia Internazionale*, XV, May 1962.
10. R.S. Bhambri, *Ibid.*
11. H. Kitamura in M.S. Wionczek (ed.) *Economic Theory and Regional Economic Integration of Asia*, New York: Praeger, London, 1966, 53.

To sum up, the foregoing arguments show the limited relevance of the trade creation/diversion criteria, as defined by the traditional theory, to problems of LDCs' economic integration schemes. These arguments suggest, though without proof, that the dynamic effects of integration are favourable to the welfare of LDCs and possibly to the world's welfare. Thus, when economic integration is viewed from the standpoint of LDCs alone, the case for their integration becomes substantially persuasive.

1(b). *Limited relevance of the traditional theory to LDCs*. Based on a number of factors, which the conventional theory has discussed in the context of DCs, some generalisation are reached to judge the desirability of economic integration. Some writers take these generalisations to apply to DCs and LDCs alike. Allen, for instance, suggested that 'although these criteria were designed specifically for application to DCs they apply to less developed areas as well.'[11] The following discussion will attempt to demonstrate just how limited is the relevance of these generalisations to economic integration among LDCs when looked at from a dynamic viewpoint. Six of these generalisations come readily to mind:

(i) Viner has raised the issue of competitiveness and complementarity in product markets and suggested that the more the partners are competitive (complementary) in the sense of producing similar (dissimilar) products, the more (less) favourable economic integration would be. Makower and Morton added that with larger cost differences among partners the gain from economic integration would also be larger.[12]

By definition, specialisation in primary products by LDCs amounts to being more competitive in the Vinerian sense; yet this general state of competitiveness, on balance, limits the welfare gain of economic integration among LDCs. This is self-contradictory, but the irony is clear. The plain fact is that most LDC exports of primary products are oriented to DC markets; consequently economic integration among LDCs, in these circumstances, would not bring about a sizeable expansion of their intra-zonal trade (see Table 2.1). However, the category of primary products is too broad, and once it is disaggregated, potential expansion would appear quite likely.

But even so, the criterion of competitiveness and complementarity is not particularly relevant to LDCs unless it is given a different

11. R. Allen, 'Integration in Less Developed Areas', *Kyklos*, 14 (1961), 315-35.
12. H. Makòwer and G. Morton, 'A Contribution Towards a Theory of Customs Union', *Economic Journal*, March 1953, 35.

sense. It presumes a developed economic structure which, when integrated, would re-adjust through a 'creative destruction' process that ends up with the survival of the most efficient producer. These economic structures are not established in LDCs, let alone the creative destruction process. Evidently, the welfare gain or loss from these effects is relevant to manufactured goods and local foodstuffs rather than to traditional exports of primary products. It must be noted, of course, that as industrialisation proceeds, the LDCs are going to be more competitive; so what they should strive for is a pattern of investment which will introduce a substantial degree of complementarity for the future.[13]

(ii) The standard theory of economic integration holds that the larger the size of the customs union, the larger the gains in welfare. If GNP is taken as a measure, one implication is that the gain from integration among LDCs is small or even negligible. Although this is understandable, a small absolute gain might in a relative sense, be quite important for LDCs. Moreover, the gain depends not only on the given size of the union but also on the rate at which it increases.

(iii) Inadequate transport facilities tend to limit the gain from economic integration among LDCs. This has been demonstrated in Chapter 2 and elsewhere. The removal of tariffs between Nigeria and Mali, for instance, would not add significantly to the market for industry established in either country. The reason is that there is at the moment no reliable direct means of surface transport between the two countries.

Transport difficulties have dealt a serious blow to effective competition and free circulation of goods in West Africa. Transport problems and, particularly, high transport costs make it difficult for the exports of the landlocked countries to compete with those of coastal states. The former are often compelled to reduce producer prices below those prevailing in the coastal countries thereby lowering rural incomes and encouraging smuggling.

In places, where transport facilities do exist, they were developed in the past with an eye to encouraging the export of primary products to the industrial countries of Western Europe and North America and are today generally inadequate for intra-regional trade. For instance, the lack of rail links between landlocked Niger and the other Entente states has affected the volume and prices of goods traded between them. Nonetheless, transport facilities should be seen only as a single parameter among many, although their

13. At present, African economies are not 'potentially very complementary'; their ratio of foreign trade to total is very high and their pre-union volume of intra-regional trade is low and in some instances zero. The sluggish performance of existing integration schemes in West Africa discussed in Chapter 5 can be largely attributed to this.

improvement should be included in evaluating the desirability of economic integration in LDCs.

(iv) It is generally believed *a priori* that the higher the initial tariff rates and the lower the common external tariff, the larger the welfare gain of economic integration. More often than not tariff rates in most LDCs are quite high either for revenue or protection purposes; hence the welfare gain would tend to increase with integration. In reality, however, there are no good grounds to expect economic integration to end up with a low common external tariff, first because the protective policy will be extended to the region, where partners can reach agreement faster if protection is increased, and secondly because it is argued that 'customs protection (even in the case of EEC) is the only effective means of securing the conditions essential to permit the co-ordination of national policies prior to their amalgamation.'[14]

It is sometimes contended that the common external tariff might have to be higher than the national tariffs of the partners. The obvious problem in this case is that it will conflict with the GATT rules whereby the external common tariff must not be higher than the initial national tariff. Nevertheless, this might be justified during the early period of economic integration among LDCs if it could be shown that successful economic integration would enable partner-countries ultimately to lower the external common tariff due to real cost reduction and improved competitiveness.

(v) It can easily be deduced from the conventional theory of integration that 'a customs union is more likely to raise welfare the higher is the proportion of trade with the country's union partner and the lower the proportion of trade with the outside world.'[15] As indicated earlier, the intra-regional trade of LDCs is small, rarely exceeding 12 per cent of their total trade (except in South-east Asia); among the EEC countries it is estimated to be well over 30 per cent.[16] The obvious implication here is that welfare gains from static effects will be small in the economic integration of LDCs. This can surely be confronted with empirical evidence.

Some empirical studies aimed at quantifying the static gains of integration among LDCs are summarised in Table 3.1. Not surprisingly, the major common finding of these studies is that the static gains of economic integration are extremely small. These gains are calculated from the standpoint of an individual country or area.

---

14. A. Marchal, 'The European Economic Community and the Developing Countries', *Annuals of Public and Co-operative Economy*, 1965, 52.
15. R. Lipsey, 'The Theory of Customs Union: a General Survey', *op. cit.*, 273.
16. See R.S. Bhambri, *op. cit.*, 236.

Even allowing for some statistical margin of error and for the inadequacy of estimates, a basic implication of these studies is that they provide no economic support for the case of economic integration. The case for LAFTA in Table III.1 bears eloquent testimony to this point.

Table III.1
STATIC GAINS OF ECONOMIC INTEGRATION AMONG EMPIRICAL EVIDENCE

Study	Case	Gain % of GNP
1. A. Singh, in H. Leibenstein, 'Allocative Efficiency vs. "X-Efficiency" ', *American Economic Review*, June 1966, 392–415.	Gains from trade among LAFTA countries, using Scitovsky method (time period not stated but it seems to be during 1961–2).	0.0075% annual growth rate of GNP.
2. P. Robson, *Economic Integration in Africa*, London: George Allen and Unwin, 1968, 91.	Gain from integration among the former members of the East African Common Market (Kenya, Tanzania and Uganda) during 1963–70.	0.5% annual growth rate of GDP or US $59.1 million.
3. Uka Ezenwe, 'Economic Integration in West Africa', unpubl. Ph.D. thesis, University of St Andrews, Scotland, 1976, 461–2.	A hypothetical estimate of integration-induced growth rate of GNP of Ghana and Entente Council States over the period, 1975–80.	0.2% annual growth rate of GNP or US $49.4 million.

However, we have to take cognizance of a set of factors which impedes the expansion of intra-zonal trade of LDCs. The most important of these factors are the low level of economic development, the inadequacy of transport facilities, over-valued currencies which cancel out significant cases of comparative advantage, foreign exchange control and other import restrictions, lack of knowledge and inadequate marketing skills, the historical ties of colonial economic integration, negative attitudes of nationalism, and the absence of standard specifications. Past experience shows that the removal of some or all of these obstacles might result in an increase in intra-regional trade. Some notable examples are the relatively high percentage of trade among the former East African Common Market members (16 per cent), and the relative increase in trade among Central American Common Market members, nearly two-thirds of whose total volume of (their) trade consists of manufactured products. Trade between Egypt and Syria after their union in 1958 showed a relatively larger increase than among any other two Arab countries.

(vi) Lastly, according to the traditional theory, 'a customs union is more likely to raise welfare the lower is the volume of foreign trade'[17] as a percentage of GNP of member-countries. The implied corollary here is that for LDCs economic integration does not promise a significant gain in welfare. However, it would appear that the relative importance of foreign trade depends upon the size more than the level of economic development. Furthermore, the relatively large volume of foreign trade represents a potential for dynamic production and income effects.

It is easy to see from the foregoing that the traditional theory of economic integration does not illuminate the structural and dynamic problems of the LDCs, since it furnishes no adequate diagnosis for evaluating the rationale of integration among them; and its generalisations are of little explanatory value. This has led some writers, like Linder, to assert that 'the possibility of a universal theory of customs union and economic development is automatically ruled out.'[18]

2. *Role of measures of policy harmonisation*

Another issue worth discussing at this point is that the traditional theory of integration does not generate satisfactory conceptualisation or testable formulas which could be applied to ensure the equitable distribution of the gains or losses of integration. In fact, it is commonplace that there is nothing in the operation of a common market to ensure the automatic distribution on an equitable basis of the gains it generates. On the contrary, it is the inherent tendency of the market mechanism to work in a disequalising manner. If we assume — as indeed this study does — that each member of a prospective or existing union is primarily concerned not with the total size of the pie but with its own share of it, there must therefore be a clear preference for 'regulated' union rather than a 'laissez-faire' union.[19]

However, measures taken with a view to correcting imbalance in the distribution of costs and benefits must not obstruct the development of the region as a whole and so become self-defeating. Besides,

---

17. R. Lipsey, 'The Theory of Customs Union: a General Survey', *op. cit.*, 273.
18. S. Linder, *op. cit.*, 32.
19. See A. Hazlewood (ed.), *African Integration and Disintegration, op. cit.*, 14.

it is important to remember that some measures necessary to deal with short-run problems may not be sufficient to correct underlying imbalances, while other measures appropriate for the latter purpose would not be useful for resolving immediate problems. Therefore exclusive pre-occupation with long-term measures should not allow short-term problems to assume crisis proportions such as would lead to the dissolution of the grouping without it having had the opportunity of tackling the long-term problems.

As we shall demonstrate in Chapter V, reluctance to accept the discipline of subordinating national interests to those of the group as a whole has been the major drawback of the existing integration arrangements in West Africa. Although Article 1 of the 1959 Convention of the West African Customs Union stipulated that members should not levy customs or fiscal duties on trade with other union members,[20] it was difficult for member-states to adhere strictly to the provisions of this Article. Against the background of the post-independence increase in government expenditure needs and the importance of duties and taxes on imports as the largest single source of government revenue in these countries, each member-country modified its fiscal duties unilaterally in accordance with its own fiscal needs.

Similarly, the partnership of unequal partners has often widened rather than narrowed the 'economic gap' between members of a grouping. Recently, many existing integrative schemes among LDCs have experienced this tendency. During the life of the West African Customs Union its activities were dominated by Ivory Coast and Senegal, a factor which contributed to its eventual collapse. While today Ivory Coast's share of intra-Entente exports is about 41 per cent, its imports account for only 10 per cent of the total.[21] In certain respects, Kenya occupied an analogous position within the East African Economic Community. Further afield, similar experiences abound. In South America, economic disparities among the members of the Latin American Free Trade Association are well known. After some two decades of co-operative regional development, *per capita* income in the region still ranged in 1979 from about $550 in Bolivia to $2,230 in Argentina and $3,120 in Venezuela. With their greater capabilities to take advantage of tariff concessions and complementarity agreements, Argentina, Brazil and Mexico have been able to increase their intra-regional trade at a faster rate than most of the less-developed members. Even in Central America, where differences in levels of development are somewhat less

20. IMF, *Surveys of African Economies*, vol. 3, Washington DC, 1970, 15.
21. See Chapter V.

pronounced than they are in South America, new industry has tended to arise in the relatively more advanced centres in El Salvador, Costa Rica and Guatemala, by passing the two less developed countries, Honduras and Nicaragua.

Thus any feasible integration arrangement likely to withstand the test of time will demand the assurance of an equitable distribution of benefits. Indeed, from the point of view of individual countries the formation of new unions or the durability of existing ones would depend largely on the prospect of gain — either directly, through faster economic growth, or indirectly, through structural transformation or both.

In a laissez-faire economic union, industry tends to concentrate in the more advanced members so as to enjoy the benefits of its larger markets and of the external economies and linkages produced by the existence of other industries and its more developed infrastructure. The lagging members suffer in more ways than one. They buy the products of the partners at a higher price than from the outside world, and they do not have the benefits which the more advanced members get in the form of higher income and employment in industry, the development and growth of external economies and the contribution which industry may make to the structural transformation of the economy.[22] For the lagging members of a union this is a serious matter.

Some of the writers discussed earlier — such as Johnson and Cooper — who have concerned themselves with the reformulation of the traditional theory of a customs union with particular reference to LDCs, have placed great emphasis on distributional considerations by building into their analytical framework some instruments for distributing gains from integration. In their conceptualisation the social objective function of industrialisation among LDCs was generally recognised, a factor which explains the difficulties involved in negotiating and launching an integration scheme, especially among countries at different levels of development. It is, of course, difficult *a priori* to generalise on the policy instruments of distribution best suited to any integration schemes since this is clearly a function of the type of integration arrangement adopted in each case.

However, there are at least four areas of policy manoeuvre: fiscal policy, monetary policy, payment and credit agreements, and industrial and investment policy. These policy instruments are discussed

---

22. P. Robson, 'The Reshaping of East African Co-operation', *East African Economic Review*, December 1967, 2. See also J.D. Cochrane *et al.*, 'LAFTA and the CACM: a Comparative Analysis of Integration in Latin America', *Journal of Developing Areas*, 8 (1), 1973, 18.

in some detail in Chapter VI in relation to the ECOWAS. Here we shall confine ourselves to some general comments intended to underline the role which the harmonisation of these policies and measures could be expected to play in the smooth running of economic groupings.

There is a case for the harmonisation of fiscal systems (including fiscal incentives) within a union on conventional efficiency grounds. *A priori*, the integration or harmonisation of the various domestic taxes and duties of an economic union makes it possible to realise the full benefits of integration and also tends to ensure a better allocation of resources, quite apart from being a useful anti-smuggling device. Lack of any measure of harmony of fiscal systems is likely to hinder intra-group trade or divert such trade to illegal channels, and ultimately lead to undesirable distortions and influences on the distribution of opportunities for productive investment.

However, fiscal harmonisation need not necessarily imply the equality of tax rates. Indeed, as Professor Dosser argues,[23] the standard tax union theory is largely inapplicable to most LDCs. He contends that the establishment of a common system of taxation for most groupings of LDCs would prove a complicated and largely meaningless exercise, given their diverse structures, principles, rates of taxation and so on, some or all of which might have to be brought into a common form.[24] Compared with the simple reduction of tariff rates of customs unions, the changes involved in a tax harmonisation programme may involve equalisation or planned differences in all these aspects. Besides, the effects of such changes need to be more widely evaluated than on the customary allocative efficiency of customs union theory, including effects on growth, balance of payments and the like. Thus, rather than embark on full tax integration or harmonisation with its attendant problems, Dosser suggests planned tax concessions in the form of rate or structural provisions as a supplement to tariff policy for chosen sectoral development, which would avoid such large and complicated questions as equalisation, trade and welfare effects.[25] In other terms, some instruments can be used in a planned way to achieve important aspects of the goals of tax integration without necessarily a recourse to complete tax harmonisation.

Similarly, members of a grouping require some degree of cooperation in their monetary and payments fields in their own

23. D. Dosser, 'Customs Unions, Tax Unions, Development Unions', Institute of Social and Economic Research, University of York, *Economics Series*, no. 145, 86–104.
24. *Ibid.*, 90–1
25. *Ibid.*, 94.

interest. The lubricating function of a common currency may be of crucial importance in the advanced stages of economic integration because a single currency allows complete freedom of payment by any one place to any other in the union. But before the stage of currency union is reached, compensation can be provided to less developed members within the framework of payments arrangements by granting more liberal credit to the less developed partners that are incurring deficits in trade with the rest of the group. However, measures in the field of payments, while useful as a temporary cushion to deficit countries in dealing with their intra-union commitments, provide only narrow scope for correcting the structural imbalances of countries not deriving their fair share of the benefits of integration. This suggests that the surplus countries should bear progressively more of the burden of balance of payments adjustment measures.

Industrial and investment policy issues differ in some respects from monetary and fiscal measures. For while the former focus mainly on preventive or *ex ante* measures designed to achieve a desirable future pattern and allocation of investment, the latter concentrates principally on correctives or *ex post* measures instituted primarily to correct existing uneven development and to strengthen the preventive measures.[26] The importance of a harmonised industrial and investment policy in an integration scheme cannot be over-emphasised in view of the quest for rapid industrialisation (from the individual country viewpoint) among LDCs. And any acceptable investment policy must guarantee each union member a fair share of potential industrial investments. Hence such a policy would necessarily involve the package approach which would be economically the most attractive and politically the most feasible. The package approach usually involves inter-governmental agreement on the establishment of designated new plants together with the necessary implementation measures. That is to say it requires the participants to come to an agreement on a specific acceptably balanced package consisting of a list of industrial projects. The main merits of this package approach as compared to the single-industry approach is that it offers the possibility for each of the co-operating countries to get a fair share in the distribution of integration-induced projects and their benefits; and, *ipso facto*, it provides a vehicle for a more equitable distribution of the fruits of integration.

26. Of course, the distinction between preventive and corrective measures is hardly clear-cut. Preventive measures should allow for great flexibility in their application, so that they can be reviewed in the course of integration, which implies, to some extent, converting preventive to corrective measures.

In this chapter, we have examined the traditional theory of integration and found its conceptualisations inadequate and largely inapplicable to LDCs, especially in West Africa. There are three main reasons for this: first, they were based on heroic static assumptions associated with advanced economies; secondly, because economic integration among LDCs should be treated as a strategy of economic development rather than as a mere tariff issue; and thirdly, because integration theory should recognise and give adequate attention to the role of protection and policy harmonisation in the economic integration of LDCs. Consequently, attempts have been made to reformulate and refine the conventional theory with special emphasis on its dynamic aspects to make it more amenable to the problems of LDCs. Great improvements have been made in this direction but almost two decades of reformulating and redefining have not yielded satisfactorily coherent conceptualisations for the evaluation of the dynamic effects of integration among LDCs. For one thing, the empirical evidence on the magnitude of the dynamic effects is far from being conclusive, being mostly related to problems other than economic integration. As Leibenstein emphasises, there are some indications of a substantial increase in labour productivity in individual firms where dynamic efficiency measures such as plant layout reorganisation, simple technical alterations, waste control, worker training and supervision were introduced.[27] Indeed, among the factors which affect the rate of growth (i.e. dynamic efficiency), Harberger has singled out technical advance to be the most important. He states: 'If there is any key factor at all for achieving rapid development, I believe it is technical advance.'[28] Thus, it is the inescapable conclusion of this review that the analysis of the dynamic aspects of economic integration among LDCs requires further empirical studies and a more systematic theoretical treatment.

27. K. Leibenstein, 'Allocative Efficiency versus "X-Efficiency" ', *American Economic Review*, June 1966, 392–415.
28. A. C. Harberger, 'Using the Resources at Hand More Effectively', *American Economic Review*, May 1959, 134–46.

IV
DISTRIBUTING THE COSTS AND BENEFITS OF INTEGRATION

The previous chapter identified the principal sources of gains or losses arising from economic integration. But it is not necessarily the magnitude of these benefits and/or costs that ensures the effectiveness and cohesion of integration arrangements. It is rather the establishment and effective implementation of a fair and acceptable distribution formula. The experience of schemes already in existence suggests that, where inadequate attention is given to the problem of equitable distribution of the fruits of integration, tensions are likely to arise. Indeed, some arrangements, especially in LDCs, have been unanimously dissolved as in the case of the Customs Union of West African States (UDEAO). Others like the Central African Customs and Economic Union (UDEAC) and the EACM have broken down, while some others, such as the LAFTA, have become for all practical purposes weak preferential trade zones which are largely ineffective umbrellas over the heads of several groups with conflicting immediate interests.[1]

We shall look now at the common causes of friction and uneven distribution of the benefits of integration which past and present schemes in West Africa have faced, with a view to suggesting some policy remedies.

1. *Sources of uneven distribution of benefits*

The two major areas of friction within an integrated scheme are those concerning, respectively, the fiscal and financial distribution arrangements and the equitable allocation of the integration industries. We consider the former first.

1(*a*). *Customs revenue effects*. One of the problems which countries entering an integration scheme have to face is the prospective loss of customs revenue. This is a matter of considerable importance to LDCs, especially those in Africa south of the Sahara where customs duties normally account for a large share of total government revenue[2] and where the raising of revenue from other sources is difficult. Indeed, since revenue duties are primarily levied on manufac-

---

1. See M.S. Wionczek (ed.), *Economic Co-operation in Latin America, Africa and Asia: a Handbook of Documents*, Boston, Mass.: MIT Press, 1969, 11.

tured goods, and where since it is the objective of integration schemes progressively to replace manufactures imported from outside the integrated market by goods produced within it, the loss of customs revenue is likely to be quite large, where a given scheme succeeds in its purpose.

The loss of customs revenue has two dimensions. It occurs both in the countries that establish industries to replace imports from outside the integrated market and in their partner-countries that import their products. Viewed from the side of the producing countries, governments will gain revenues — albeit small ones — from the taxes paid by the producing firm and the manpower it employs, which, it is hoped, will compensate in varying degrees for the loss of customs receipts, whereas the governments of the importing countries might not have offsetting increments in tax receipts.[3] Also from the standpoint of the intra-union importing partners, trade diversion quite often means not only loss of customs revenue but, at least in the short run, more expensive goods than those of third countries. Because the weaker importing partner-countries would pay higher prices for similar goods hitherto imported from third countries, the balance of trade and payments of these countries would consequently be worsened. Of course, each member of an integration scheme will eventually become both a producer of exports to its partners and an importer of the latter's exports, so that each government will have some additional tax receipts from enterprises supplying the sub-regional market to compensate for losses of customs revenue. Even so, the more developed members will tend to become net exporters of manufactures to their less developed partners and they will be able to offset losses of customs receipts by additional tax revenues from productive activities to a greater extent than the less developed member-countries. Polarisation explains this.

---

2. The share of import duties to total revenue among ECOWAS members ranges from around 20 per cent in Ghana to about 70 per cent in Togo with the other countries falling somewhere in between. Nigeria, where import duties contribute only about 5 per cent, is the only notable exception; the petroleum sector alone furnishes over 85 per cent of public revenue (see IMF, *Surveys of African Economies*, vols. 3 and 6). Thus the revenue loss effect of integration for the overwhelming majority of ECOWAS members could be quite serious.

3. Extra direct tax receipts may not be large enough to offset the revenue loss from customs duties. Even so, there is no reason why union members, who feel badly hit, should not invoke Article 26 (1) of the ECOWAS Treaty to introduce substitute revenue duties (such as consumption and sales taxes) on a non-discriminatory basis (i.e. levied equally on domestic production, imports from partner countries and imports from third countries). See UN, *Current Problems of Economic Integration: The Distribution of Benefits and Costs of Integration among Developing Countries*, New York, 1973 (TD/B/394), 17.

Polarisation or 'backwash' effects have their basis in economies of scale in infrastructure, external economies in the labour market and elsewhere, and mutual support of market demand created in advanced sub-regions, when trade throughout the region, even when free of tariffs, is subject to transport costs.[4] The story of international common markets among LDCs is replete with cases of polarisation. The reshaping of the now defunct EACM in 1967 was precipitated largely by disputes arising from Kenya's attraction of a proportionately greater share of industries. Within the Arab Common Market Egypt occupies an analogous position. The Central American Common Market has been partly affected by polarising growth. In this case, the most advanced country, in terms of income levels, has not been the most rapid in its growth. But if Costa Rica has been growing more slowly than Guatemala, Nicaragua, and El Salvador, all four of these countries have greatly widened the gap between their levels of income and that of the poorest country and the one with the slowest growth, Honduras.

A more general survey of regional inequality in national development by Williamson, to which Edel refers,[5] indicates that increasing disparities are customary throughout at least early phases of development. Thus while regional inequality becomes stable and eventually falls in the more developed countries, for countries at the West African level of development the net effect of economic integration on regional economic growth might be to increase disparities between the member-states, at least until deliberate offsetting provisions are instituted.

Let us illustrate this point from ECOWAS. Members of ECOWAS should be considered as falling into two broad categories — the poorer and the more advanced — based on some mixture of the 1975 estimates of the value of GNP, *per capita* income and size of the domestic market as an index of measure (Table II.1). The more advanced members would comprise Nigeria, Ghana, Ivory Coast, Senegal, Sierra Leone and Liberia while the rest would belong to the relatively poor group.

The poorer members would need special concessions over the freeing of trade. Articles 13 and 14 of the ECOWAS Treaty[6] stipulate a two-year tariff grace period, an eight-year tariff establishment span

---

4. See Uka Ezenwe, 'Economic Integration in West Africa', unpubl. Ph.D. thesis, University of St Andrews, Scotland, 1976, 475.

5. See M. Edel, 'Regional Integration and Income Redistribution: Complements or Substitutes?' in R. Hilton (ed.), *The Movement Toward Latin America Unity*, New York: Praeger, 1969, 185-99.

6. See Treaty of the Economic Community of West African States (ECOWAS), Lagos, 1975.

and a further five-year common external tariff establishment period for all members. Although the fifteen-year span of progressive tariff disarmament and the establishment of a common external tariff for the entire Community appears reasonable, the Treaty failed to make any special provision for the weaker members. Instead of requiring all members to 'progressively reduce and ultimately eliminate import duties' immediately following the initial two-year period, when members may not be required to reduce or eliminate duties, we recommend that this provision be re-considered with a view to granting the poorer members a minimum of twelve years within which to eliminate duties on intra-ECOWAS trade. Thus they will have four years longer than the advanced countries to eliminate tariff on intra-union trade.[7] When the issues of intra-union tariff disarmament have been solved, the entire union could then work progressively towards the establishment of a common external tariff.

As for the problem of uneven distribution of the loss of customs revenue, a form of compensatory tax can be devised which we shall call 'off-setting tax' (OT). Both in feature and orientation this will be very similar to the *taxe unique* which was introduced in the UDEAC or the transfer tax in the EACM which came into operation in 1967. The *taxe unique*, which is tantamount to an excise duty, was collected on finished products at the source of production in return for the exemption of manufactures from import duties on raw materials and equipment coming from third countries.[8] The OT, like the *taxe unique*, would be at a considerably lower rate than the import duties and taxes on equivalent imports, and the proceeds from this excise tax would be transferred to the treasuries of member-countries consuming products of regional origin covered by this fiscal arrangement according to the volume of their imports.

The principal aim of the OT is to compensate for a fall in fiscal revenue resulting from the elimination of duties on goods from intra-union sources and to distribute the revenue fairly;[9] but it is

---

7. This kind of concession is not uncommon. Under the sub-regional Agreement of the Andean Common Market, Bolivia and Ecuador, which are relatively underdeveloped, enjoy special treatment and concessions. See W.P. Avery and J.D. Cochrane, 'Sub-regional Integration in Latin America: the Andean Common Market', *Journal of Common Market Studies*, December 1972, 85–102.

8. M.S. Wionczek (ed.), *Economic Co-operation in Latin America, Africa and Asia*, Boston, Mass.: MIT Press, 1969, 235.

9. The type of compensation for the loss of revenue envisaged in Article 25 of the Treaty of ECOWAS is unlikely to be satisfactory. It is usually difficult to assess correctly the amount of such losses. Besides, funds are hardly available in the right amounts to pay adequate compensation to member-states which have suffered loss of import duties as a result of the application of the Treaty.

capable of achieving two other important goals: to encourage the local manufacture of consumer goods and to ensure sufficient control over exemptions granted on imports of raw materials or equipment used in local manufacture while encouraging the infant industries concerned.

The efficient operation of the OT would require close administrative co-operation among ECOWAS members in the field of tax collection. Also some co-ordination of investment policies would be desirable to avoid serious distortions in conditions of production. A common investment regime would not only provide for the incorporation into the OT scheme of regional enterprises needing access to the regional market but it would also provide for the co-ordination of foreign investment regulations which contain differences in incentives offered in different parts of the region. Such co-ordination would not necessarily imply uniformity, as differences in the incentives offered to foreign investors might be desirable in order to offset the relative disadvantages of some locations over others and to help a flow of investment into the partner-countries in accordance with the regional strategy of diffused development. However, such mutually agreed differences must be sharply distinguished from differences based on historical accident or reflecting unilateral competitive decisions of the partners to secure investments at each other's expense (see Articles 29, 30 and 31 of the ECOWAS Treaty).

In a way the application of the OT arrangement should be seen as a medium-term solution. In the very long run it would be advisable on practical grounds for the governments of ECOWAS member-states to consider a general reform of their tax systems with a view to reducing their dependence on revenue duties through the introduction of a system of consumption taxes. Unquestionably, this will eventually become necessary even in the absence of an integration scheme, since national import-substitution policies would, in any event, reduce the importance of customs duties as a source of government revenue.

1(b). *Allocation of integration industries.* The other problem area relates to the allocation of integration industries. This problem has two principal aspects. The first has to do with a fair distribution of the benefits generated by *existing* regional industries, while the other concerns an orderly reduction of the disparities between the economies of members through the application of an agreed formula for distributing *future* regional industries. 'Existing regional industries' are industries which existed prior to integration, but which needed a wider market than was available domestically to operate at full capacity or optimally. The OT, as we suggested earlier, can take care of the problem of distributing the benefits generated by existing integration industries. The more thorny question, to which we now turn, is the allocation of future regional industries.

A fair and equitable distribution package deal should incorporate a system of allocation of future industries. This would normally involve assigning virtual monopoly rights — say, by licence — for the production of certain industrial commodities to the favoured country. Experience suggests that in order to make the assignment of industries to member-countries easier and more acceptable, it has to be based on an agreed list of specified industries.[10] The principal rationale for any assignment of monopoly rights lies in the presumed existence of economies of scale. Given this assumption, it can be asserted that an outright allocation of monopoly rights will be economically more profitable than allocation through the market mechanism, since the latter implies the appearance of many smaller industries which would gradually drive each other out of existence until the largest surviving one would be the only supplier in the market.

The benefits to be derived from the allocation of industries will depend primarily on the care with which the allocation is made; that is to say, if the industries are allocated to the countries which are 'more efficient' in the respective products, then presumably the gains to the union as a whole will be maximised. If the industries are allocated to their worst producers, then the gains will be substantially reduced. The fundamental questions therefore are how to distribute the industries between member-countries so as to maximise the gains to the union, and how to distribute the benefits to be derived from the union in a 'fair' manner.

The application of purely economic principles such as location theory, comparative cost and economies of scale is unlikely to produce a sufficient measure of agreement to solve the problem. Because members of ECOWAS may have differing notions about equity and accord varying degrees of social value to the apparent cost of sub-optimal production in determining their national political strategies, negotiators will take into account particular objectives and wishes of individual ECOWAS countries before arriving at the final allocation system. For instance, some may be willing to 'trade-off' employment against income, or income in one part of the country against income in another part of the country.

Of course, such problems as inter-state differences in employment opportunities can be taken care of through the implementation of Article 27 which permits intra-ECOWAS mobility of labour. But the

---

10. See M.S. Wionczek (ed.), *op. cit.*, 184.

institution of intra-union mobility of labour to compensate for interstate differences in employment opportunities will at best be a short-term solution. In the long run, any acceptable allocation arrangement will guarantee each member a 'fair' share of the new industries.

In general, any distributionally constrained allocation will imply an efficiency loss, for industries will not always be installed where they produce the highest benefit for the group as a whole. A 'trade-off' solution is therefore called for here. For the sake of stability and success of any scheme, a sub-optimal allocation may be acceptable, but a skewed distribution of the benefits to members would be intolerable.

2. *Institutional arrangements necessary to implement a distribution package*

A regulated distribution system, such as we have suggested, involves by its very nature an interventionist approach, which inevitably requires permanent institutional arrangements for its enforcement. Institutions appropriate for a given integration scheme would depend largely on such factors as the type of integration scheme in operation, the range of products involved, the anticipated value of net benefits (benefits minus costs) *vis-à-vis* the administrative costs, the locational pattern of integration industries, the impact of intra-union trade on individual members' total external trade and the political feasibility of any chosen institutions.

Chapters 2 and 11 of the ECOWAS Treaty provide for a considerable number of institutions and commissions charged with different functions. Of immediate interest for the implementation of our distribution system are two organs; the Fund for Co-operation, Compensation and Development (FCCD) and the Industry, Agriculture and Natural Resources Commission (IANRC). The administration of OT could be assigned to the FCCD. Since the OT principle involves a measure of income transfers from the 'winners' to the 'losers' in the regulated distribution sense, the FCCD can physically collect the refunds and make payments out of its receipts to the net losers. This would surely go a long way towards eliminating the periodic crises which have bedevilled members of groupings where the operational form of compensation have involved direct income transfers of one kind or another from the richer to the poorer partners.

As for the allocation of future industries with a view to influencing the distribution of their impact on members' economies, the IANRC would be the appropriate body to carry out this function. Because of the lack of information required to quantify the effects of future industries by means of projections, any distribution measure taken *ex ante facto* would have to be on the basis of somewhat speculative estimates. Hence, such a system will have to allow for great flexibility in its application, so that it can be reviewed in the course of integration.

In the particular case under review we advocate a system of industrial licensing. This system, as noted earlier, will prevent ruinous competition and operate as a monopoly concession for the enterprise that obtained the licence and for the country in which it is situated. But to ensure an accelerated and orderly diffused industrial development of the area through the application of the industrial licensing system, other supportive measures would be necessary. There should be well-articulated guidelines for the granting of licences, including an agreed list of industries, and these guidelines should be biased in favour of diffused regional development. More important, there should be an incentive scheme to back up the granting of licences.

What emerges from the foregoing is that a common market may not further economic equality among its members unless an equitable distribution mechanism is set up to ensure it. This presupposes that an integration arrangement will generate net benefits both to the group as a whole and to each one of the individual member. By implication, it means that each member of the group must be at least as well-off as a member as it would be if it had not joined the scheme. Needless to say, the computation of these benefits would be easier under a regime of common pricing policy for regional enterprises; and the only way to ensure the full co-operation and continued membership of each partner is to guarantee it a fair share of the net benefits of integration.

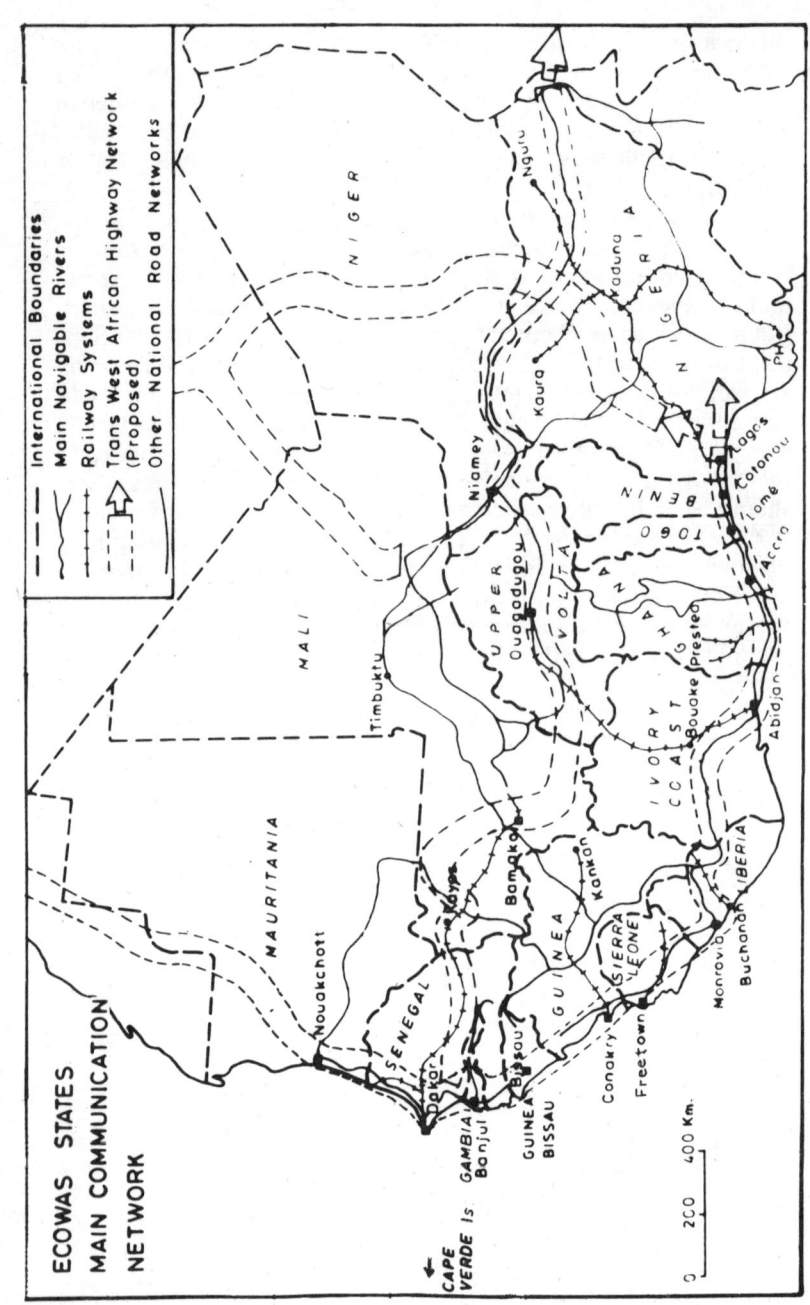

V
THE PERFORMANCE OF EXISTING INTEGRATION SCHEMES IN WEST AFRICA

1. *General*

Despite the adverse legacies of colonialism, as demonstrated in Chapter I, several sporadic efforts have been made on economic co-operation in West Africa in more recent years, especially by the Francophone countries,[1] to build on the foundation of past economic co-operation. These attempts have had the blessing since 1964 of the UN Economic Commission for Africa, in co-operation with the Organisation of African Unity, because both organisations have come to regard economic co-operation as a pragmatic way of overcoming, or at least mitigating, the adverse economic effects of narrow national markets. There are currently at least twenty economic co-operation schemes either defunct or functioning or in a state of paralysis in the sub-region (Chart V.1). These organisations include those of which the territorial spread covers the whole of Africa as well as those that cannot be strictly defined as economic integration arrangements. Nevertheless, they all contain, in their orientation and approach, elements of politico-economic co-operation. Since their existence has some bearing on the process of economic integration in the area, they deserve notice here.

West Africa as a geopolitical expression presents a formidable challenge to received integration theory and practice. We noted earlier that the conventional theory classifies patterns of integration into one or more of the following multinational economic formations: free trade area; customs union; common market; economic and supranational union. But this traditional compartmentalisation does not suit the West African experience. Indeed, most of the

---

1. As noted elsewhere, Francophone West Africa has advanced much farther in this respect than Anglophone West Africa. This was partly because the Francophone territories, unlike the Anglophone countries, had the advantage of geographical contiguity; but a more important explanation is perhaps that the French empire, of which French West Africa was a part, was more closely knit economically than the British empire to which the countries of British West Africa belonged. Economic co-operation among British West African countries took the form of a common currency, the West African pound and the sharing of a number of technical, research and commercial institutions, such as the West Africa Cocoa Research Institute and the West African Airways Corporation. But after independence each country set up its own institutions.

Table V.1. WEST AFRICA: MULTILATERAL ECONOMIC ORGANISATIONS

	ECOWAS	UDEAO	CEAO	UMOA	Entente	OERS	Senegam	Mano	CBC	RNC	WACH	ECA	ADB/ADF	OCAM	CPA	IACO	APC	OCCMED	Lomecon	OAU
Benin	×	×	×	×	×					×	×	×	×	×		×		×	×	×
Cape Verde	×												×						×	×
Gambia	×						×				×	×	×				×		×	×
Ghana	×										×	×	×		×				×	×
Guinea-Bissau	×											×	×						×	×
Guinea	×					×				×		×	×			×			×	×
Ivory Coast	×	×	×	×	×					×	×	×	×	×	×	×		×	×	×
Liberia	×							×			×	×	×					×	×	×
Mali	×	×	×	×		×				×		×	×	×					×	×
Mauritania	×	×	×	×		×			×		×	×	×	×		×		×	×	×
Niger	×	×	×	×	×				×	×	×	×	×			×	×		×	×
Nigeria	×									×	×	×	×			×	×	×	×	×
Senegal	×	×	×	×		×	×				×	×	×	×			×	×	×	×
Sierra Leone	×							×					×						×	×
Togo	×	×	×	×	×						×	×	×		×	×		×	×	×
Upper Volta	×	×	×	×	×					×	×	×	×	×				×	×	×

KEY TO ACRONYMS AND ABBREVIATIONS IN TABLE V.1

Intra-West African

ECOWAS	Economic Community of West African States
UDEAO*	West African Customs Union
CEAO*	West African Economic Community
UMOA*	West African Monetary Union
Entente	The Entente Council
OERS	Organisation of Senegal River States
Senegam	Senegambian Integration Scheme
Mano	Mano River Union
CBC	(Lake) Chad Basin Commission
RNC	River Niger Commission
WACH	West African Clearing House

Extra-West African

ECA	(UN) Economic Commission for Africa
ADB	African Development Bank
ADF	African Development Fund
OCAM	Organisation Commune Africaine et Malagache
CPA	Cocoa Producers' Alliance
IACO	Inter-African Coffee Council
APC	African Groundnut (Peanut) Council
OCCMED	Organisation of Co-ordination and Co-operation for the Fight against the Major Endemic Diseases
Lomecon	Lomé Convention
OAU	Organisation of African Unity.

*For original French versions of these acronyms refer index.

organisations presented in Chart V.1 are hardly subject to any of the customary classifications, if only because none limits co-operation to trade matters only.

To be able to elicit the logic and rationale behind the establishment and operation of these organisations, a new scheme attempts a different classification system.[2] According to this, relations among co-operating states are classified as homogamic (integration exists among neighbouring states); anabatic (among states non-neighbouring but homogamically related by proxy); satellitic (between developed central and underdeveloped peripheral states); and stochastic (a random residual system of relations among states). Although still not very satisfactory, the new classification system provides a more meaningful framework for categorising the multilateral economic formations in West Africa.

While it is not intended here to examine the historical background of each of these integrative movements — except for the more important intra-West African groupings — in order to determine their appropriate group, a quick glance at Table V.2 will reveal that most of the organisations come under homogamic and anabatic groupings. The groupings formed exclusively by the Anglophones or the Francophones or both can be classified as homogamic, provided that the members occupy a contiguous geographical area. As Table V.2 shows, these would include ECOWAS, Mano Union, CBC, OERS and Senegam. There are several others which are partly homogamic and partly anabatic, especially those involving homogamic Francophone groupings which are anabatically separated in the

Table V.2
CLASSIFICATION OF MULTILATERAL ECONOMIC ORGANISATIONS IN WEST AFRICA

Homogamic	Homogamic/Anabatic	Satellitic	Stochastic
ECOWAS	UDEAO	Lomecon	
Mano Union	CEAO		ADB
CBC	UMOA	ADF	
OERS	IACO		ECA
Senegam	Entente		
	RNC		
	WACH		
	CPA		
	OCCMED		
	OCAN		
	OAU		

2. See G.C. Abangwu, 'Towards West African Economic Unity', *Africa*, no. 36, August 1974.

geographical sense by non-member Anglophones. Such hybrid homogamic-cum-anabatic organisations are UDEAO, CEAO, UMOA, IACO, Entente, RNC, WACH, CPA, OCCMED, OCAM and the OAU.

The remaining integrative movements are easier to classify. The signatories of the Lomé Convention, for example, epitomise a case of satellitic relationship both in character and orientation. For, as discussed in Chapter II, it is a relationship between the rich centre represented by the European members of the EEC and the poor periphery represented by the ACP member-states. As regards the ADF, it also qualifies as a satellite body. Though an affiliate of the ADB, the ADF is largely funded by non-African members, mostly European and American countries, which contributed, as at 31 December 1977, 97 per cent of the Fund's capital and commanded no less than half of its voting strength; it has also commanded one-third of the remaining half since 1978 by virtue of their subscription to the ADB's capital of one-third of the total and their exercise of voting power equivalent to this proportion of the total.[3]

Indeed, by allowing non-African countries to subscribe to the Bank's capital in 1978, the Board of Governors has changed the Bank's character. It is no longer an exclusive African affair but a stochastic organisation with homogamic, anabatic and satellitic elements. The ECA belongs to the same class; a United Nations body, it is run by staff provided by the Secretary-General of the United Nations based in New York. Naturally, the staff consult and obtain approval for their actions from New York. At the same time the Commission, which is usually headed by an African Executive Secretary, is based in Addis Ababa, the capital of Ethiopia, and charged with responsibility for African economic and social problems. Be that as it may, that these organisations are satellitic or stochastic by themselves does not make them any less important and beneficial than other organisations. What really matters is their relevance and contributions to the development of the region.

The foregoing brief comment on the form of existing schemes in the region serves to draw attention to their nature, problems and possible impact on economic integration in West Africa. For the rest of this chapter we will discuss the experiences and performance of the most important intra-West African groupings.

---

3. See K.D. Fordwor, *The Annual Report* of the African Development Bank and African Development Fund, Abidjan, 1977; and Central Bank of Nigeria, *Annual Report*, 1978, 91.

2. The Customs Union of West African States (UDEAO)

2(a). *Structure and organisation.* The former members of the French West African Federation (Afrique Occidentale Française), having learnt from their common history the benefits of corporate development, decided as they were approaching independence to form a customs union. The only dissenting voice was Guinea. On 9 June 1959, the West African states of Benin (Dahomey), Ivory Coast, Mali, Mauritania, Niger, Senegal and Upper Volta signed a convention which established the first customs union of West African states (Union Douanière entre les Etats de L'Afrique Occidentale (UDAO). These seven countries intended to maintain the customs union regime in effect before independence and to harmonise import taxation among themselves. Essentially, the UDAO provided for the absence of customs duties and quantitative restrictions on goods in trade between its members but did not call for the elimination of other duties and taxes.

By the standard definition[4] of a customs union, the UDAO had many essential characteristics to qualify as one. The absence of tariff and non-tariff barriers between the partners, the establishment of a common external tariff and the general harmonisation of tax legislation in member-countries were all provided for. The three principles[5] which governed the union can be outlined as follows:

(i) The convention stipulated that UDAO members should not levy customs or fiscal duties on trade with other members (Article 1). This provision has to be viewed against the background that duties and taxes on imports have been by far the largest single source of government revenue in these countries, as indeed in most LDCs, partly because foreign trade can be taxed more easily than other sectors and partly because direct taxes are difficult to collect in economies with relatively limited monetised sectors and weak fiscal machinery. Between 1964 and 1966 average proceeds from duties and taxes on imports accounted for about 60 per cent of total tax in Benin, Ivory Coast, and Upper Volta; 45 per cent in Senegal; 40 per cent in Mali; 34 per cent in Nigeria; and 33 per cent in Mauritania (Table V.3). Over the same period, the ratio of revenue from duties and taxes on imports to the value of total imports was estimated at 47 per cent in Upper Volta, 39 per cent in Benin, 35 per cent in Ivory

---

4. It means the suppression of discrimination in the field of commodity movements within the union and equalisation of tariffs in trade with non-member countries. See Bela Balassa, *The Theory of Economic Integration, op. cit.*, p.2.

5. See J.E. David, 'La création d'une Organisation de Co-opération Industrielle, Economique et Douanière entre les Etats membres (UDEAO) in IMF', *Surveys of African Economics, op. cit.*, 14–44.

Coast, Mali and Senegal, 28 per cent in Niger, and 27 per cent in Mauritania.

(ii) The establishment of a common external tariff extended to both customs and fiscal duties levied by the member-countries (i.e. imports from non-member-countries were made subject to a common customs tariff). The system, as noted earlier, involved the adoption of a two-tier tariff structure. But, although customs duties were levied for fiscal as well as protective purposes, they were not applicable to imports from France. And they also varied between goods imported from countries receiving most-favoured-nation treatment (goods subject to the minimum tariff) and goods imported from other countries (goods subject to general tariff, usually three times as high as the minimum tariff). Fiscal duties in effect on 31 March 1959 and applied to imports from non-member countries were to be maintained, and all parties to the convention were to agree on any modification or introduction of new duties. The convention specifically precluded any member from changing the level or structure of its import taxes without prior concurrence of all other partners (Article 1).

(iii) The UDAO Convention provided for a Committee of Experts to study the harmonisation of the laws and regulations on internal taxes in order to prevent illicit trade between the member-countries (Article 3). This committee was also entrusted with undertaking studies on other matters designed to facilitate the implementation of the convention, particularly problems of double taxation and tax evasion. Measures dealing with such problems could be embodied in multilateral or bilateral agreements to be worked out by the interested parties. The convention did not specifically call for coordination of similar investment regulations; but this objective was partly implied in the expected harmonisation of the fiscal systems.

The UDAO Convention also provided for the establishment of several common institutions with a view to settling problems arising from the implementation of the customs union (Articles 2–6). Apart from the joint commissions which were to determine the distribution of revenue from import and export duties and taxes and apart from the Committee of Experts which (as noted above) was to study fiscal harmonisation, the convention established a Customs Union Committee vested with decision-making powers (Article 5) and composed of one representative from each member-state, its decision to be binding on all parties. The Customs Union Committee was to make decisions on important matters, such as the distribution of revenue collected from import and export duties and taxes and the proposals for fiscal harmonisation submitted by the Committee of

Experts. Disputes arising from implementation of the customs union could be brought before the Court of Arbitration of the French Community.[6]

Thus on paper it seemed as if the UDAO would be very successful. But that was not to be. As we explain below the Convention virtually became inoperative and inoperable in its original form shortly after it had come into force.

2(*b*). *Performance.* The in-built weaknesses of the UDAO Convention of 1959 made its orderly implementation extremely difficult. Its goals were over-ambitious, its provisions were inflexible and its machinery was too complicated to ensure smooth and effective implementation. We shall consider some of the salient weaknesses of the convention.

First, the system of revenue allocation was complex and dilatory. The convention states that receipts from imports and export taxation should be distributed in such a manner as to give each country its appropriate share (Article 2). Joint commissions were to be set up in two or several member states for determining the distribution of receipts on the basis of customs declarations, investigations among traders, and other relevant elements of estimation. Disbursement of receipts was to be made quarterly. To deal with this, arrangements were worked out between member countries for a system of sharing proceeds from duties and taxes on imports. Senegal and Mauritania had an arrangement whereby such proceeds were distributed on the basis of 91.38 per cent for the former and 8.62 per cent for the later. Similar arrangements existed between Ivory Coast and Upper Volta and between Benin and Niger. Since the 1959 Convention did not settle the problem of origin of goods or define the goods originating within the UDAO area, and whereas goods originating in non-member countries were subject to taxation and reimbursement each time they crossed country boundaries within the UDAO area, the administration of the revenue allocation system proved cumbrous and woolly. There was also the additional factor that various regulations infringed upon the importers' freedom to choose points of entry, and hence forced them to distribute their imports according to countries of final destination.

Secondly, the Convention was very flexible over the right of individual member states to alter its tax structure. Given the fast growth of government expenditures in the post-independence era, it was difficult for member states of the UDAO to adhere strictly to the provision of Article 1 which forbade them to change or modify their

6. IMF, *Ibid.*

The Customs Union of West African States (UDEAO)

import tax system to meet their revenue requirements without the concurrence of all the other members. In practice, each member-country did modify its fiscal duties in accordance with its own financial needs. The base of fiscal duties was broadened, rates were raised, and excise taxes were increased. Besides, products originating in member countries were subjected to fiscal duties, beginning in 1962.

Thirdly, the Convention failed to provide for adequate consultation between the member-states, with the result that no progress was made in the field of fiscal harmonisation. The single body vested with collective decision-making, the Customs Union Committee, seemed to be too supranational and powerful to allow for a flexible and realistic approach. Unlike the Equatorial Customs Union established in the same year, UDAO did not call for varied and permanent institutions such as a general secretariat, an executive committee and a council of heads of state or ministers. In the absence of adequate permanent institutions, none of the goals of the 1959 Convention was achieved, except for the establishment of a common customs tariff in all UDAO countries.[7]

Even so, except between Senegal and Mauritania, whose trade links have in any case been particularly close, the impact of the Convention on the flow of intra-regional trade had been extremely limited. In the face of this disappointing performance, the UDAO members met in Paris in 1966 to seek ways and means of reshaping their near-inoperable union. Opinions came down heavily in favour of restructuring the organisation; it was hoped that transforming it into a looser association would make it more effective.

2(c). *Reshaping of the UDAO.* The result of the Paris Conference was the signing at Abidjan in 1966 of a new convention, establishing with effect from 15 December 1966 the Customs Union of West African States (Union Douanière des Etats de l'Afrique de l'Ouest, or UDEAO).[8] Taking into account the diverse fiscal interests of the different states, the UDEAO was more realistic than the UDAO. Indeed, the reshaped union was more of a free trade area than a customs union. However, it is ironical that while the UDEAO countries transformed their customs union into a broader economic union in 1966, the West African (UDEAO) countries were at the same time going in the opposite direction by transforming their customs union into a much looser customs grouping.

---

7. *Ibid.* As the next chapter will show, permanent institutions play a vital role in the smooth and effective operation of integration schemes.
8. *Ibid.*

Table V.3
UDEAO COUNTRIES: RELATIVE IMPORTANCE OF RECEIPTS FROM IMPORT DUTIES[1] AND TAXES,[2] 1964–66
Import duties[1] and taxes[2] as % of . . .

	. . . Tax Receipts			. . . Imports		
	1964	1965	1966	1964	1965	1966
Benin	60.4	66.0	57.9	37.0	38.6	40.3
Ivory Coast	59.7	62.0	63.4	34.0	34.5	37.3
Mali	51.0	35.6	37.6	36.3	29.1	41.4
Mauritania	41.2	30.2	29.5	36.8	21.3	22.8
Niger	27.1	36.3	28.7	30.6	26.1	26.4
Senegal	47.2	46.2	44.1	37.8	35.7	35.0
Upper Volta	60.5	63.8	n.a.	45.8	50.0	n.a.

Source: IMF, *Surveys of African Economics*, vol. 3, Washington DC, 1970, 15.

Notes: 1. Fiscal and customs duties. 2. Import turnover tax, and statistical and other surcharges on imports.

In terms of scope, as noted above, the 1966 Convention did not provide a complete customs union. It sought to ensure the free circulation of local products of members within the area, subject to a certain measure of protection for existing industries and a fiscal tax. To facilitate this, the Convention provided for a common customs duty; it also called for a further harmonisation of legislation on other import and export taxation in recognition of the evolution of tariff legislation in each country. It excluded any tariff concessions below the minimum tariff on goods imported from countries outside UDEAO (Article 3), but allowed for concessions between the minimum and the general tariff, and the origin of imported goods was defined. In a nutshell, the new agreement was more modest, less imprecise and more realistic in its provisions.

The organisational set-up was also more conventional, although this had also led to bureaucracy. A number of permanent institutions consisting of a Council of Ministers, a Committee of Experts, and a General Secretariat were established. The Council of Ministers was the supreme policy-making body. The council met once a year in ordinary session and was made up of one member of each government who was ordinarily the Minister of Finance. Its decisions required a simple majority and were binding on all members. Beneath this was a Committee of Experts which acted as an advisory organ of the Secretariat while the General Secretary, based in Ouagadougou (Upper Volta), conducted the day-to-day operations of UDEAO under the Council of Ministers.

2(d). *Impact of the UDEAO on trade.* Within the intra-union area, commodities grown, extracted or manufactured were exempted from customs duty, but were subject to fiscal duties and other taxes, which in total should not exceed 50 per cent of the aggregate of the most favourable treatment accorded to similar goods imported from third countries (Article 6). A higher rate amounting to 70 per cent was authorised to protect an industry that might be less competitive than a new, similar industry in another member-country.[9] The Convention even allowed pre-existing bilateral arrangements among member-countries: for example, the agreements between Ivory Coast and Upper Volta, signed on 19 March 1963, and between Senegal and Upper Volta, signed on 7 September 1965, remained operative when the UDEAO came into being. Although these agreements were judged to be compatible with the provisions of the 1966 convention, the UDEAO scarcely justified the high hopes of its founders. The volume of trade between the member-states from 1966 through 1969 increased by 1 per cent for imports and 2 per cent for exports.[10] (It ceased to function on 1 June 1972.[11])

Several reasons can be adduced to explain this phenomenon. Generally, the economies of the former UDEAO members do not fulfil the conditions of the traditional theory of integration. They are not 'potentially very complementary'; their ratio of foreign trade to the total, as was shown earlier, is very high; their pre-union volume of intra-regional trade was low and in some instances zero while the scope of economies of scale was limited, especially where intra-union protective tariffs were conceded.

Planned complementarity seems to be a necessary process in the integration of developing economies. If industries are to be efficient, not only must they be assured of access to a wider market but they must also be specialised on the basis of factor endowments and availability of raw materials. Thus, allowing for their scarcity, resources should be utilised so as to maximise comparative

---

9. It is interesting to note the similarity between these provisions and those accorded to Tanzania under the Kampala Treaty of 1967 (see P. Robson, 'The Reshaping of the East African Co-operation', *East African Economic Review*, December 1967). Although this necessary safeguard is designed to diffuse development rather than encourage polarisation within the union, it makes for trade-diversion. In terms of pure theory, it is counter-productive because consumers in the protected infant industry area within the union pay higher prices for similar goods associated with the new industries than the consumers in the other parts of the common market where concessional protective duties are not enforced. This may, of course, be transient and could be defended on dynamic-efficiency grounds, namely revenue, employment and industrialisation.
10. C. Legum and Associate (eds), *Africa Contemporary Record*, 1969–70, C448.
11. *African Research Bulletin*, 2375A.

advantages. Duplication of productive units could be avoided through harmonisation of industrial development within the area and reconciliation of national development plans. Without complementarity, the volume of intra-union trade would continue to be low, since governments would often invoke safeguard clauses or take fiscal measures to protect industries from their competitors in other member countries. Furthermore, given the present pattern of production and foreign-oriented trade, it is difficult to expand intra-union trade despite the existence of the customs union. But if, within the framework of the modernisation and diversification of agriculture, market-oriented food production had been expanded and larger import-substitution industries had been established within the UDEAO, the proper implementation of the customs union might have increased intra-union trade considerably. In other words, the expansion of the intra-union market would have made it easier to realise the economies of scale.

Another visible problem, which bedevilled the UDEAO and contributed more than any other single factor to its collapse,[12] is the question of disparities in development levels. Ivory Coast and Senegal, the most advanced countries in the area, have been developing much faster than the landlocked countries. Available statistics monitor this 'development gap' very clearly: 44 per cent of manufacturing units established in the area originate from Dakar and Abidjan, both of which also account for 60 per cent of electrical power generated and consumed for industrial use. The share of Senegal and Ivory Coast in UDEAO's foreign trade was of the order of 75 per cent in 1965–7.[13] Thus the UDEAO's inability to cope with the problem of sharing its benefits and its tendency to polarise development to the disadvantage of the less developed countries must have contributed to its downfall. However, as we noted in the last chapter, the problem of polarisation of development is nothing new to integration literature. If there had been a strong political will among members to overcome this problem, the introduction of some equitable distribution measures could have saved the situation. There is much too to learn in this connection from the EAEC experience.[14]

---

12. See *African Research Bulletin*, 2375A.   13. IMF, *op. cit.*, p.58.
14. The EAEC had its fair share of a similar problem. The problem of disparity of benefits of the Economic Community has persisted. In 1961 the Raisman Commission recommended introduction of a measure of fiscal compensation from Kenya to Tanzania and Uganda but this did not remove Tanzania's dissatisfaction. The Kampala-Mbale Agreement of 1965 represented a further attempt to meet Tanzania's claims and to some extent Uganda's by three chief means: a selective system of firm production reallocation in favour of Tanzania; the imposition on an

The Customs Union of West African States (UDEAO)

Finally, transport difficulties dealt a serious blow to effective competition and free circulation of goods within the UDEAO. These problems, discussed elsewhere, and in particular high transport costs made it difficult for the exports of the inland countries to compete with those of coastal countries. The former were compelled to reduce producer prices below those prevailing in the coastal countries and thus lowered rural incomes. At the same time, the landlocked countries paid higher prices for their imported goods, compared with prices in the coastal areas. This anomalous situation points to the urgent need for an adequate integrated transport network in West Africa.

2(e). *The final collapse of the UDEAO.* On 11 May 1972, the organisation's Secretary-General Mr Tamboura said in Dakar that the UDEAO would cease to function with effect from 1 June and would be replaced by the Communauté Economique de l'Afrique de l'Ouest (CEAO).[15] He argued that the three principles of the Customs Union — free exchange of goods and persons between member-states, the establishment of a common external customs tariff, and the sharing of customs receipts amongst all member-states — had not been respected by member-countries, 'faced for the most part with budgetary problems'. In the light of this situation, he went on, the Heads of State decided on 3 June 1966 to set up a new Convention which would encourage the member-states to respect

---

agreed basis of some quantitative restrictions on inter-country trade; and a formula for the allocation of new regionally-oriented industries so that Tanzania would get the most. This Agreement was not fully implemented, and so was ineffective. Partly as a consequence, Tanzania resorted in 1965 and 1966 to sweeping unilateral import restrictions against Kenya.

In the light of this development, the Philip Commission was set up to examine and recommend ways of strengthening the EAEC and of diffusing development, particularly industrial growth, within the Community. It recommended, *inter alia*, decentralisation of the headquarters, quantitative restrictions on basic staple foods or major export crops subject to special marketing arrangements, and imposition of the so-called transfer tax. The recommendations were embodied in the 1967 Kampala Treaty and their implementation greatly assuaged the fears of Tanzania and Uganda (See P. Robson, *The East African Economic Review*, December, 1967).

Similar developments affected the Central American Common Market (CACM). Honduras suspended common market trading in 1970 when its neighbours Guatemala and El Salvador, which had attracted most of the industries in the region, refused to do anything to help her attract new factories. In September 1972, the representatives of Guatemala, El Salvador and Nicaragua announced they had closed their borders to goods originating from Costa Rica; the action came after Costa Rica decided to impose dual exchange rates, a measure her neighbours claim amounts to a 30 per cent import surtax (see *New York Times News Service*, 8 September 1972). The three remaining members agreed to limit trade among themselves to current levels.

15. *African Research Bulletin*, 2347A (see Section 5.2).

UDEAO's principles. But this decision did not apparently materialise since the Heads of State meeting in Bamako in 1970 gave the general secretariat the task of replacing UDEAO by another organisation having a wider area of co-operation objectives.

On 3 June a treaty instituting the CEAO was signed by the Ivory Coast, Benin, Upper Volta, Mali, Mauritania, Niger and Senegal at Bamako. Togo stayed out, although it sent an observer. Mr Tamboura described the new body as an 'instrument of progress in the service of our peoples'.[16] But as we shall soon see in our later discussion of the ECOWAS and CEAO, this remark is no more than an *obiter dictum*.

Thus it can be seen that, even before the final collapse of the UDEAO, it had become obvious to observers that the grouping needed thoroughgoing reorganisation and reorientation, if anything was to be salvaged from the union. In this regard, mention should be made here of J.E. David who made a careful study of the UDEAO's problems[17] containing a number of interesting suggestions as to how to re-orient the organisation into an effective organ of corporate development. David envisaged the creation of an organised zone of commercial exchanges, based on common interests and good faith, among those member-states of UDEAO that felt they would on balance benefit from membership and that would work hard to ensure the success of the integration scheme. According to David the conclusion of inter-state agreements, which would ultimately lead to a customs union, would form the initial framework of co-operation. These agreements, however, were to include the adoption and definition of the rules of origin and a tariff structure conducive to intra-union trade, the proceeds from which would go to a distributable pool. Furthermore, there was to be the creation of an 'encouragement fund for industrialization and for inter-state commerce'.

The merits of David's model are clear. The UDEAO might have collapsed in any case, but the spirit and momentum of economic co-operation in the region could not only have been kept alive but also been given a new lease of life. The emerging groupings might have been smaller but manageable. And the integrating economies might have been at sufficiently similar levels of development for none to be in a position to dominate others. However, now that the UDEAO is no more, it will be interesting to see how far David's ideas influenced the formation of the succeeding grouping (see Section 3).

16. *Ibid.*
17. J.E. David, *op. cit.*

The West African Economic Community (CEAO)

We have shown how the UDAO of 1959 became a restructured body in 1966 in a bid to throw off its in effectiveness. But although the reshaped and renamed UDEAO was very modest both in orientation and temper, it never fulfilled the aspirations of its members. The difficulties facing the organisation, as we have demonstrated, were essentially structural and organisational; the revenue need of individual states exacerbated by the disparities in inter-country problems, especially those of landlocked members, impaired intra-union free flow of goods and services; and lastly, inadequate and ineffective administrative machinery made the enforcement of trade regulations difficult.

Even if David's proposals for a radical reorientation of economic co-operation in the area were accepted, some of the key obstacles to integration would still persist. This is because these problems (non-complementarity of primary production and trade, public revenue problems and transport difficulties) are fundamentally structural by nature, and are likely to face any future integration experiment in the region. It can therefore be said, without mincing words, that the logic of the situation demands a frontal attack on these structural obstacles.

3. *The West African Economic Community (Communauté Economique de l'Afrique de l'Ouest (CEAO)*

Francophone West Africa's newest grouping, the Communauté Economique de l'Afrique de l'Ouest (CEAO), was formed to replace the defunct UDEAO. As noted in the preceding section, a preliminary agreement instituting the CEAO was signed at Bamako on 3 June 1972. But the CEAO was not formally established until 17 April 1973 following a meeting of heads of state in Abidjan, where the final treaty was signed. The Treaty of Abidjan came into effect on 1 January 1974.

3(a). *Membership, structure and organisation.*
The CEAO has six members: Ivory Coast, Mali, Mauritania, Niger, Senegal and Upper Volta. These are the French-speaking states of West Africa which signed the final treaty. Although Benin signed the preliminary Bamako accord, it refused to sign the final treaty in Abidjan, opting instead for observer status. Togo signed none of the agreements. The new organisation was intended to combine the Entente with the Organisation of Senegal River States (OERS). With the exception of Guinea and Benin, it covers what was until 1957 French West Africa.

CEAO aims to create a free-trading zone among the Francophone

West African countries. It is clearly based on the franc zone and on Yaoundé-type association with the EEC. Its stated aims are to harmonise and boost the economic activities of its members.[18] Emphasis is placed on the expansion of trade between members, particularly in agricultural and industrial products, and to this end it provides for the joint promotion of agricultural and industrial development and the co-ordination of transport and communications. There are special protocols on livestock production, fishing, the compilation of statistics and accountancy methods. A common export policy is envisaged, and complex internal trade arrangements are to be worked out.

The CEAO is not a thoroughgoing customs union, which would inevitably have encouraged lopsided development within the union, as the UDEAO actually did. Hence there are to be special preferences, designed — like the imposition of a transfer tax under the EAEC Treaty of 1967 — to minimise the cumulative polarisation effects of intra-union free trade on the less developed members. Also the CEAO Treaty provides for the establishment of a development fund to compensate member-countries for any revenue losses sustained through the substitution of CEAO-made goods for extra-union imports. The organisation's headquarters at Ouagadougou will administer the fund.

Compared with the UDEAO, the CEAO is clearly less ambitious in orientation and more realistic in its structure. It is smaller and looks more manageable. It would appear that J.E. David's idea of a smaller, more manageable zone of commercial exchanges based on common interests and good faith influenced the founding fathers of the new organisation.

Should the CEAO actually succeed in creating a common market for the franc-zone countries of West Africa, it could become a more close-knit, better balanced and more effective economic integration scheme than its predecessor, the UDEAO. To ensure its success, the Treaty of Abidjan established four main bodies and a series of organs responsible for implementing joint policies, and some advisory committees. The four main bodies are: the Conference of Heads of State, the Council of Ministers, the Secretariat and the Court of Arbitration. The Conference of Heads of State, the Community's highest authority, meets at least once a year and its decisions have to be unanimous. The Council of Ministers, consisting of the responsible ministers of all the member-states, meets at least twice a year and is responsible for promoting any action conducive to achievement of the Community's objectives. The

---

18. See *Africa Confidential*, 11 May 1973, and *West Africa*, 17 December 1973.

Secretariat, headed by a Secretary-General appointed by the Conference of Heads of State, is responsible for preparing and implementing the decisions of the Conference of Heads of State and of the Council of Ministers, provides Secretarial services to those bodies, and carries out studies on their behalf. As for the Court of Arbitration, it is charged with the responsibility of solving disputes which may arise between states of the Community regarding interpretation of the Treaty and the protocols thereto.

3(b). *CEAO problems.* The CEAO's future is still very uncertain. Its problems are many and multifarious, and are not only economic but also political.

In the first place, the creation of the CEAO is part of a complex manoeuvre to counterbalance Nigeria's political and economic weight, and in particular to check her attempts to organise a larger West African economic community in which the former French colonies might achieve a greater degree of politico-economic independence from France. But recent thinking on economic integration in West Africa does not favour the CEAO spirit.

Internally, the cohesion of Francophone Africa seems to be crumbling (and with it some of France's influence). As pointed out earlier, Benin and Togo dissociated themselves from the CEAO by refusing to sign the Abidjan Treaty. Benin argues that it has commitments to its neighbours and principal trading-partners Nigeria, Ghana and Togo and since none of these belong to the CEAO, it would be unrealistic to join it. Togo, which has close ethnic and economic associations with Ghana, has been more interested in working with Nigeria towards the formation of a wider West African economic community, that would embrace both Anglophone and Francophone countries, than in the CEAO. Indeed, President Eyadema of Togo and General Gowon, then head of state of Nigeria, agreed in April 1972 to form the 'nucleus' of such an integration scheme.[19] It is widely believed that this was an attempt by Nigeria to get in ahead before the purely Francophone formation was announced. It is possible that the CEAO was discussed and planned during the first African tour of the late President Pompidou of France in 1971 when he advised Francophone countries to counter-balance the heavy weight of Nigeria.[20] Nigeria's reaction to this took the form of an intensified campaign for a wider West African grouping — which resulted in the signing of the ECOWAS Treaty, discussed in Chapter VI.

19. *Ibid.*
20. *Ibid.*

Within the CEAO itself some members have expressed strong reservations.[21] The erstwhile President of Niger and first CEAO chairman, Hamani Diori, was quoted as saying that it would be very 'unrealistic' to set up a West African Economic Community without the participation of English-speaking countries, arguing that Nigeria was Niger's main African trading partner. Even Mauritania had misgivings and wanted to reserve the right to extend to her Maghreb neighbours the same preferences she is to enjoy in the CEAO. Besides, her decision to have her own currency impeded the franc-zone arrangements with the CEAO.

Externally, too, the CEAO got off to a bad start. The CEAO members declared officially that their new grouping was open to 'all African states, be they Francophone or Anglophone . . . to join us and adhere to the Treaty'.[22] But what is actually meant here is that other African countries could, if they wished, join the CEAO — once it was established — in the future. The point is unmistakably implied in a remark, associated with President Houphouet-Boigny, on the absence of Anglophone representatives at the Bamako Conference: 'Great Britain did not take part in the establishment of the EEC and has only later asked to join.'[23] To say the least, the analogy being drawn here represents a serious misjudgement. The weight of Nigeria relative to the membership of the CEAO is enormously greater than that of the United Kingdom relative to the EEC. The entire population of the CEAO (including Benin and Togo, which have so far refused to join) is less than 30 million, less than half the population of Nigeria. The GNP of the CEAO countries, except for the Ivory Coast and Senegal, is pitifully small, as is their economic growth rate. In 1978, the combined GNP of the present CEAO members accounted for only 31.7 per cent of that of Nigeria.

Geographically, the Francophone CEAO is hardly a unit. As already noted, Nigeria is already Niger's major trading partner, while Ghana separates Togo and Benin from the other members of the CEAO and has strong economic ties with Togo. If anything, the circumstances of West Africa demand greater horizontal economic integration than at present. Thus it is hard to justify the importation of Anglophone-versus-Francophone relations, as they exist in

---

21. See *Africa Confidential*, 11 May 1973.
22. *West Africa*, 7 July 1972. This statement is credited to President Houphouet-Boigny of Ivory Coast.
23. *Ibid*. There can be little doubt as to the intention to keep the Anglophones out of the CEAO. Ghana applied to attend the Bamako meeting but was told by Senghor it was too late — the intention was to invite the Anglophones to join 'later'.

Europe, into the fundamentally different local circumstance of West Africa.

Observers of the West African scene doubt whether the setting up of an exclusive Francophone grouping offers the right solution to the problems of the sub-region. More recently, however, there has evidently been some rethinking on this matter. At their meeting in April 1975, the heads of state of the six member-countries invited the other countries of West Africa to join the organisation on the founders' terms.[24] This invitation of course was not accepted by the Anglophones — and the founding fathers of the CEAO, Senghor and Houphouet-Boigny, probably knew that it would not be.

The only published result of the two-day closed session at Niamey was the launching of the existing community, largely ineffective since its formation in 1973, into the 'operational phase' of Francophone co-operation and the election of President Senghor as the first Chairman of the CEAO.[25] But, even after nine years of existence, the future of the community still appears uncertain. What is clear is that its formation has brought to the surface some suppressed differences among the Francophones. For the first time some Francophone countries, in the face of strong French encouragement, if not pressure, opposed the creation of yet another exclusive organisation of Francophones which would tend to perpetuate the pre-independence cleavage. Furthermore, the traditional conflict of interests between the relatively rich coastal countries and the impoverished, semi-desert landlocked states of the interior was brought frankly into the open. For instance, exports entering the intra-zonal market from Senegal and Ivory Coast currently account for over 70 per cent of the total.

Given these internal differences, the attainment of economic union through linguistic groups does not adequately take account of the stark realities of the sub-region. The signing of the ECOWAS treaty by all the sixteen states of West Africa in Lagos surely indicates that a persuasive case for a wider and all-embracing grouping in the area still exists, despite the existence of the CEAO. To some extent, the founding of the CEAO marks yet another milestone in Houphouet-Boigny's long quest for an inter-state organisation vehicle suited to his leadership.

4. *West African Monetary Union (UMOA)*

Perhaps, the most successful post-AOF arrangement for economic co-operation in West Africa has been in the monetary field.

24. *West Africa*, 5 May 1975.
25. *Ibid.*

Although in some respects the West African Monetary Union (Union Monétaire Ouest Africaine — UMOA) would be regarded as the monetary arm of the UDEAO, the former, unlike the latter, has stood the test of time, despite the defection of Mali.[26]

On 12 May 1962 a treaty establishing UMOA and providing for a common currency and a common central bank was signed by the member-states of the UDEAO. Togo joined later in 1963. As far back as 1959, the UDEAO members had shared a central bank, the Banque Centrale des Etats de l'Afrique de l'Ouest (BCEAO), which issued a common currency, the CFA franc. However, the old BCEAO was dissolved on 31 October 1962 and was replaced by a new central bank of the same name but with enlarged responsibilities and an intergovernmental organisation. On the same day the BCEAO countries concluded with France a co-operation agreement whereby France guaranteed the convertibility of the CFA franc, issued by the BCEAO, into French francs, and the members of the UMOA undertook to keep their external reserves in an operations account opened by the new BCEAO at the French Treasury, with which a special relationship was established. It must be noted at this point that the treaty establishing the UMOA and the agreement for co-operation are two separate documents; indeed the treaty providing for a monetary union and a common central bank could continue in force even if the agreement for co-operation were to be abrogated.[27]

The BCEAO has a closely knit but simple organisation run at two levels. Although with its headquarters in Paris, the central bank maintains an agency in the capital of each member-country and has established sub-agencies in some other places within the UMOA territory. The overall management of BCEAO is entrusted to a board of directors to which each member-country has two appointees, from among whom one is elected president; and France, under the provisions of the co-operation agreement, appoints seven directors, or the equivalent of half of the total number appointed by the UMOA countries. As a general rule, the decisions of the board are taken by a simple majority, but in practice certain important decisions must be adopted by a two-thirds majority — indeed,

---

26. Mali did not ratify the treaty establishing UMOA, nor did it join in the co-operation agreement. However, after the 1967 devaluation of the Mali franc and negotiations with France, Mali established an operations account, with effect from 29 March 1968, and its agreements with France provide that it may eventually join UMAO (see IMF, *op. cit.*, 24). See also *Africa Research Bulletin*, 2375A.

27. IMF, *ibid.*, 71. Two new agreements, of 4 December 1973 between BCEAO and France and of 14 November 1974 between the UMOA members, replaced the earlier agreements without much changes.

amendments of the statutes require the unanimous decision of the board.[28]

Below the board, a five-member National Monetary Committee, appointed in each member-country by the government and including the two national directors, implements the general credit and rediscount policy decisions taken by the board of directors. Day-to-day running of BCEAO is entrusted to a director-general, appointed for an indefinite term by the board of directors. The director-general attends, either personally or through a delegate, all meetings of the board and the National Monetary Committees; he also represents BCEAO in all its external relations. All BCEAO personnel are appointed and removed from office by him, although appointment of agency managers requires the prior approval of the government of the country in which the agency is established.[29]

In general, the division of responsibilities between these two levels of administration is that the board, for its part, is responsible for fixing the overall supply of short-term credit in the light of resources and needs, for fixing the discount rate, and for determining the ceilings to be granted by the local branches of the Central Bank to each economy in respect of rediscounts, advances and so on. The National Monetary Committees are responsible for advising the board from time to time on the credit limits in the individual BCEAO countries and, when these have been fixed, for determining the ceilings for each local bank and for individual enterprises. Also the Central Bank can sometimes delegate some power to the Monetary Committees to act on its behalf in certain matters.

Three most important functions of the BCEAO can be distinguished. These include holding the monopoly of note issue, being a depository of external reserves and credit creation.

BCEAO, in its capacity as a Central Bank, has the sole right to issue the CFA franc in each member-country. As at 14 August 1972, the official rate of the CFA franc — which, as noted earlier, is fully convertible into French francs — was CFA Fr. 613.25 = £1.00 sterling; but the relationship of the CFA franc to the French franc has always remained[30] at CFA Fr.1 = Fr.0.02 (Fr.1.0 = CFA Fr.50). The effectiveness of UMOA in overcoming internal payment problems has been remarkable. Because notes and coins issued by BCEAO are legal tender in all member-countries and circulate freely

---

28. *Ibid.*
29. *Ibid.*
30. The apparent constancy in the rate of exchange between the CFA franc and the French franc is explained by the fact that the CFA franc is, by agreement, automatically convertible into the French franc, hence they appreciate or depreciate in value simultaneously.

within UMOA, statistics based on issuance of each country's notes and withdrawal of notes of other countries reveal a fairly large inter-country circulation of notes, particularly between Mauritania and Senegal, between Benin and Togo, and within the three states Ivory Coast, Niger and Upper Volta.[31]

The other function of BCEAO is that it acts as the depository of UMOA members' external reserves. These reserves are held in what is termed an operations account at the French Treasury. As indicated above, the procedural aspects of the centralisation of BCEAO's reserves in the operations account are regulated by a convention between BCEAO and the French Treasury. An amendment to this convention dated 2 June 1967 makes it possible for BCEAO to invest part of its exchange reserves in certain types of negotiable bonds, maturing within two years, issued by international financial institutions of which all BCEAO countries are members.[32] Since this amendment, BCEAO has invested part of its foreign reserves in short-term bonds issued by the International Bank for Reconstruction and Development (World Bank).

Pursuant to the co-operation agreement concerning the operations account, BCEAO can apply specific measures in the case of a continuous and sizeable reduction in exchange reserves. For example, BCEAO's discount rate and charges on advances must be increased by 1 per cent if the operations account for the area as a whole with the French Treasury shows a debit for sixty days.[33] Also, in the event of the deposits in this account being exhausted, BCEAO may require public and private organisations to surrender their French francs or other foreign currency holdings to it[34] against CFA francs. But BCEAO, at its discretion, may restrict this requirement solely to public institutions and banks and may implement it only in countries whose external transactions through the operations account show a deficit.

The third major role of BCEAO relates to credit creation. Within

---

31. *Ibid*, 73. Although all the notes and coins circulating within UMOA are issued by BCEAO, identification of CFA notes by a letter following the serial number enables BCEAO to keep separate accounts for each country's currency in circulation. Coins are not identified by country and they account for roughly only 4 per cent of total currency in circulation (*ibid*.).
32. *Ibid*.
33. *Ibid*., p.74.
34. It must be stressed here that the French Treasury pays interest to BCEAO on balances in the operations account at an annual rate at least equal to the rediscount rate of the Bank of France, but never less than 2.5 per cent. In 1967/8 the effective rate was 5.3 per cent. The BCEAO, on the other hand, pays interest to France on any overdraft balance at an annual rate between 1 per cent and the rediscount rate of the Bank of France, depending on the amount (see *ibid*.).

the framework of rediscount ceilings which are the principal means of control, BCEAO is authorised to extend both short-term and medium-term credit. Short-term credit is extended in the form of rediscount of short-term paper and temporary advances (*prises en pension*) against private and government paper, as well as direct advances secured by either gold or foreign exchange and securities acceptable to BCEAO. Normally the period for which short-term credit is granted is limited to six months, but it may be extended up to nine months for financing crops and public contracts.[35] Medium-term credits are granted by BCEAO for periods not exceeding five years.

Furthermore, BCEAO may grant the Treasury of any UMOA country ways-and-means or short-term advances for a period not exceeding 240 days, consecutive or not, per calendar year, and in an amount not exceeding 10 per cent of the government's fiscal receipts during the preceding budgetary year. On 10 December 1968, however, the BCEAO statutes were amended to enable the board of directors, after reviewing developments in the currency issue and evaluating the effects of its decision on the development of the currency issue, to raise the maximum amount of short-term advances to an amount equal to 15 per cent of fiscal receipts.[36] In addition, BCEAO may discount Treasury customs duty bills, provided such bills have a maturity of less than four months, have been issued by a solvent debtor, and are guaranteed by a bank. Also, Treasury bills of UMOA member-countries with a maturity of less than six months may be either rediscounted or accepted for temporary advances; accepted as collateral for an advance within the limits fixed by the board of directors; or bought from or sold to the banks without endorsement, provided that banks do not act as intermediaries for the Treasuries.[37]

Despite these credit facilities, the BCEAO's provision of credits has in general been on a modest scale. Although almost all UMOA governments have at one stage or another taken advantage of the BCEAO's ways-and-means advances, only Benin, Niger and Upper Volta can be said to have made really intensive use of the facility, since they have experienced the most severe fiscal problems. Similarly, short-term credits in the form of advances secured by either gold or foreign exchange have not been provided so far.

Apart from technical factors, one probable reason for the limited use of the credit facilities of BCEAO is connected with the fact that

---

35. *Ibid.*, 75.
36. *Ibid.*, 77.
37. *Ibid.*

UMOA member-governments have often been able to obtain any needed short-term advances from the French Treasury at reasonably low rates of interest. The French attitude regarding financial support to its former colonies is relatively permissive. As the Jeanneney Report states: 'France in effect renounces the possibility of refusing to finance initiatives taken unilaterally by African governments, and in return the states accept a certain monetary tutelage, particularly in the matter of deficit financing'.[38] The corollary here is that France would continue, subject to its own financial limitations, to finance development projects from those governments which appear to it to be sound.

Also it must be recognised that the primary objective of BCEAO's credit policy is to facilitate orderly economic growth in member-countries while maintaining an appropriate balance between financial and real resources. Given this policy constraint, BCEAO's credit expansion must necessarily be kept in constant check. It is against this background that the recorded average annual rates of credit expansion — of 17 per cent in 1962–4, of 6 per cent in 1964–6 and of 13 per cent between 1967 and 1969[39] — could be regarded as reasonable.

Aside from the three principal functions discussed above, BCEAO has a variety of other responsibilities usually associated with a central bank. They include the supervision of the national credit institutions in accordance with the national regulations, managing the accounts of national treasuries, giving expert advice to the treasuries and maintaining deposits for the commercial banks.

By conventional definition, UMOA is a complete monetary union. Not only are there a complete pooling of monetary reserves and the issue of a common currency for the participating countries, but there is also a substantial integration of their financial markets and all obstacles to internal payments and transfers are removed. In this way intra-regional balance-of-payments problems are eliminated,[40] while extra-regional payments problems are alleviated since union members can add to their foreign currency holdings that portion of their foreign exchange which would otherwise have been spent in settling intra-regional trade accounts. Furthermore, UMOA through the operations of BCEAO facilitates the co-ordination of other economic policies of its members in regard to trade and

---

38. Cited in P. Robson, *Economic Integration in Africa*, 207.
39. IMF, *op. cit.*, 84–9.
40. Although the volume of intra-union trade among the UMOA members has not fundamentally improved over the past decade, nevertheless the existence of a common currency points the way towards the expansion of intra-zonal trade.

West African Monetary Union (UMOA)

Table V.4
BCEAO COUNTRIES: NET FOREIGN EXCHANGE HOLDINGS,
1962–69
(in billions of CFA Francs; end of year)

	1962	1963	1964	1965	1966	1967	1968	1969 (Sept.)
Central Bank (net)[1]								
Benin	2.47	2.45	2.44	2.51	2.27	1.98	2.35	2.18
Ivory Coast	8.67	9.79	9.34	14.62	14.92	16.97	18.21	15.57
Mauritania	1.96	2.32	2.57	2.41	1.92	2.17	1.84	1.44
Niger	2.31	2.10	1.75	0.73	0.94	0.15	0.58	1.78
Senegal	17.97	11.83	8.41	8.25	11.27	8.44	3.64	4.17
Togo	2.12	2.22	2.83	4.32	4.58	5.46	6.22	6.41
Upper Volta	3.38	3.52	3.30	3.44	4.01	4.50	5.58	5.95
Total Central Bank	40.49	35.94	32.55	38.28	42.28	42.37	42.24	45.02

1. Aggregate net holdings of the Central Bank do not equal the sum of the shares allocated to each member country because small amounts of the Bank's assets are allocable to any particular UMOA member.
Source: IMF, *ibid.*, 122.

economic development in general. These are the great advantages of a monetary union. Ironically, these advantages are also a source of weakness. In the first place, a monetary union of the UMOA variety robs its members of the power to pursue different monetary policies which their different economic problems might warrant. In particular, the limitation on the freedom of individual members to alter their competitive positions through exchange-rate manipulation *vis-à-vis* other members and the outside world can hurt the weaker union partners. Indeed, specific actions may sometimes be required to deal with the problems of such disadvantaged members. Thus, unless all members are content to keep in line with each other in all aspects of monetary policy — credit expansion, exchange control measures and so on — problems are bound to arise. Unhappily, while some UMOA member-economies are doing well, others are not; hence the latter group may not be completely satisfied with union-inspired common policies which do not allow for individual initiative. Consider the figures in Table V.4, from which it is evident that in 1962–8, while the net assets of Ivory Coast rose steadily (except in 1964), those of Senegal showed a steady decline except in 1966. The asset positions of Togo and Upper Volta also steadily improved, while those of Benin, Mauritania and Niger declined. Perhaps the freedom to adopt the necessary monetary measures, in a given situation, could have helped the reserve-scarce UMOA countries to improve their reserve positions. Thus one is bound to weigh the immense advantages of UMOA against the likely pressures which would be brought to bear upon it by the seemingly divergent development policies of different member-countries and by the effects of such policies.

However, there are at present two strong stabilising factors at work within the UMOA. The first, as we noted above, is the existence of alternative sources of finance from the French Government outside the union, thereby mitigating the foreign exchange pressure. The second factor is the flexibility and vigilance, as exemplified in the 1968 amendment, which the BCEAO has demonstrated in its credit operation. As pointed out already, this amendment increases from 10 to 15 per cent (of each member's fiscal receipts) the maximum amount of short-term advances which each member can received from BCEAO. This no doubt shows that BCEAO watches closely the credit problems of its members in their process of development and is willing to accommodate them within its own limitations.

One can, in general, doubt the ability of a monetary union to meet the credit requirements of all its members, given the possibility of differences in levels and rates of economic development among its members and the internal pressures which these differences are likely to generate. But when viewed from the standpoint of a particular case, as in that of the UMOA, and the stabilising factors at work within it, one is tempted to conclude that the advantages outweigh the disadvantages.

5. The Council of the Entente States (Conseil de l'Entente)

5(a). *Structure and Orientation.* During the days of AOF up to the formation of the UDAO, Ivory Coast discovered to its chagrin that it was contributing about half the region's exports and tax revenue but was receiving only a quarter of its imports and total expenditures.[41] The Ivorian Government was thus unwilling to continue subsidising the weaker economies of the area, especially that of Senegal, and was also anxious to counterbalance the then Mali Federation's appeal to Benin and Niger and to protect its potential markets in the other countries and commercial transport links with them. It was apparently to give expression to these aspirations that President Houphouet-Boigny spearheaded the founding of the Entente Council in 1959.[42] When the Entente was set up, Niger presented the fewest problems for Houphouet, for its then President, Hamani Diori, was

---

41. See R.H. Green and Associate, *Unity or Poverty?*, op. cit., 151.
42. *Ibid.*

one of his most devoted followers. But Benin and Upper Volta proved far more difficult. Indeed, it took all Houphouet's skill as well as financial inducements to bring Upper Volta and Benin into the Entente under the Ivory Coast leadership. The foundation members thus included the Ivory Coast, Niger, Upper Volta and Benin. Togo did not join until 8 June 1966.

In essence, the Entente is an 'informal instrument'[43] for the coordination of politico-economic policies based solidly on self-interest. Because of the similarity of their problems and of their backgrounds as former French dependencies, its members have over the past twenty years developed a good deal of understanding and a spirit of corporate existence among themselves. The Entente hardly fits into any formal definition of an integration scheme, if judged by the conventional definitions discussed above. Nevertheless, the 1959 agreement by which it was established contained two potentially important economic provisions.[44] The first envisaged the harmonisation of the countries' development plans, but lamentably this has so far remained a pipe-dream. The second provided for the establishment of a Solidarity Fund, which is discussed below.

By its very nature, the organisation of the Entente Council is loose and personalised. The heads of state and ministers of its members meet regularly, at least twice a year, during which policy decisions are taken. There is also a permanent secretariat to execute the decisions taken at the ministerial level. In practice, however, much depends on how the creator and uncontested leader of the Entente, President Houphouet-Boigny of Ivory Coast, reacts to these decisions. It is contended that his consistent moral and financial support could be regarded as the major reason for the Entente's survival to date.[45] And we shall see in the course of this discussion that there is much to support this view.

5(*b*). *History and Activities of the Entente.* As indicated above, the establishment of a Solidarity Fund was one of the key provisions in the Entente Agreement of 1959. Although Ivory Coast has always insisted on a free hand in promoting private — mainly foreign

---

43. See P. Robson, *Economic Integration in Africa, op. cit.*, 246. Also see Virginia Thompson, *West Africa's Council of the Entente*, Ithaca, NY: Cornell University Press, 1972, 274.
44. R. Robson, *op. cit.*
45. See T. Golan, 'What role for the U.S. in Africa?', in *West Africa*, 23 October 1972. Thompson asserts that their chronic and imperative need for the funds which Ivory Coast has supplied is what has kept Upper Volta, Benin, Togo, and Niger members of the Entente, for as individual countries they are not economically viable (V. Thompson, *op. cit.*, 277).

— investment in industry, the setting up of the Solidarity Fund in 1959 was *prima facie* a tacit acknowledgement on its part of the fact that unbalanced regional development, especially within a grouping, has its dangers.

On paper the fund was to receive for redistribution 10 per cent of the revenues of each member-state. But in practice members did not strictly adhere to this principle. Indeed, the Solidarity Fund seems to have operated informally as the means through which Ivory Coast (or rather its President) has subsidised the budgets of the other members as and when required, from resources not subject to public scrutiny in Ivory Coast.[46] Hence no provision for payment of the subsidies ever appeared in the Ivory Coast budget, although until recently Solidarity Fund receipts were recorded in the budgets of the other members. The accounts of the Fund were not subject to audit.[47]

However, despite the operation of the Solidarity Fund, the disparity between the wealthy Ivory Coast and its four indigent partners continued to grow. One indicator of this overwhelming economic superiority is presented in Table V.5. The Table clearly shows that Ivory Coast dominates the Entente market even when Ghana is brought into the picture.[48] While the former contributed about 41 per cent of the Ghana-Entente exports, its imports account for a mere 17.5 per cent of total. Apart from the Ivory Coast, the landlocked states of Upper Volta and Niger contribute more than the rest to intra-area trade. Their share of exports is roughly 16 per cent each while the import figures are 13.5 per cent for Niger but more than 35 per cent for Upper Volta. As indicated earlier, this is a reflection of their heavy dependence on intra-zonal trade rather than their economic strength. Ghana buys almost as much as it sells to the area. This is also true of Togo, but to a less extent. Benin exports less (8.6 per cent) and imports more (14.6 per cent).

Aside from its share, the composition of Ivory Coast's trade is another source of discomfort to its partners. It exports predominantly manufactures but imports mainly raw materials and animal products; hence Ivory Coast has been accused of promoting its own economic development by treating its partners as suppliers of raw materials and markets for its output.[49]

To this end, a further attempt was made in 1965 to infuse new life into the Entente. This took the form of the Houphouet-Boigny

---

46. See P. Robson, *Economic Integration in Africa, op. cit.*, 247.
47. *Ibid.*
48. Ghana is not a member of the Entente but it has been included in the table for comparative reasons as well as to show its close trade links with the Entente states.
49. See V. Thompson, *op. cit.*, 272.

Convention on dual nationality and economic harmonisation.[50]

The economic harmonisation provisions reflected the pressures which had been building up all along, from weaker Entente members, to encourage a more equitable distribution of industrial development. As for the reciprocal citizenship provisions within the Entente, these were primarily inspired by the large size of the immigrant labour force from other Entente countries in the Ivory Coast. However, as events came to show, the double nationality proposal proved an error of judgement on the part of the Ivorian President. He misjudged the reactions of his countrymen to the proposal, particularly the white-collar workers who bitterly opposed it on the grounds that it presented a threat to their employment. Equally, the proposal was not very popular among other Entente states, especially Upper Volta. For given the attraction which Ivory Coast offers as an island of prosperity and opportunities in a sea of poverty and deprivation, immigrants from other Entente countries would tend to settle in it. This in turn would deprive the poorer Entente states of an important element in their balance of payments account — receipts in the form of transfers from their nationals working in other Entente countries, particularly Ivory Coast. In the end the President was unable to achieve even some form of *modus vivendi*, and to his dismay was forced by the *Bureau Politique* to withdraw his proposal. The entire convention failed to obtain ratification.[51]

In the continuing search for ways and means of strengthening the Entente Council, a Mutual Aid and Guarantee Fund was created in 1966, formally superseding the Solidarity Fund, with the object of stimulating private investment within the Entente area.[52] A secretariat under Mr Paul Kaya, aided by four advisers (three French and one American), administers the Fund. But policy decisions are normally taken at the occasional meetings of the Management Committee of the Fund. As with the old Solidarity Fund, Ivory Coast has been its principal source of finance. Of its capital of CFA Fr.1,300 million, Ivory Coast provided CFA Fr.1,000 million.[53] Despite its vastly disproportionate contribution to the Fund, Ivory Coast — in order to enable the less developed members of the grouping to reap the initial benefits — agreed not to draw on the

---

50. See *Convention on Dual Nationality and Economic Harmonization*, Abidjan, 1965.
51. See the *Financial Times*, 8 December 1971. Also see P. Robson, *Economic Integration in Africa, op. cit.*
52. *Ibid*.
53. See T. Golan in *West Africa*, 23 October 1972. Also see *Entente Africaine* (a quarterly review), (6) March 1971, 7.

Table V.5
INTRA-GHANA-ENTENTE TRADE, 1968–72
(average %)

Importing countries	Exporting countries						
	Ghana	Ivory Coast	Niger	Upper Volta	Dahomey	Togo	Total
Ghana	—	1.1	1.5	3.9	0.5	3.5	10.5
Ivory Coast	—	—	4.4	10.9	2.2	..	17.5
Niger	1.9	6.0	—	1.1	3.6	0.9	13.5
Upper Volta	4.9	29.6	0.6	—	0.1	0.4	35.6
Dahomey	0.3	2.7	8.5	0.1	—	3.0	14.6
Togo	3.5	1.6	0.8	0.4	2.2	—	8.5
Total	10.6	41.0	15.8	16.4	8.6	7.8	100

Notes: Figures are subject to rounding error. Ghana is not a member of the Entente but included in the table for comparative purposes.
Source: Computed from IMF, *Direction of Trade: Annual*, 1968–72.

fund for the first five years of its operation.

The financial operations of the fund are organised at two distinct, though not mutually exclusive, levels. The first consists of the loan activities and the second relates to the management of the Fund's Intervention Budget. By the end of 1969 the Fund's loan disbursements to its members had reached a total of 1,249 CFAF million.[54] Of this figure CFA Fr.97.5 million went to the Chamber of Commerce of Upper Volta for the construction of storage facilities for its export products at Ouagadougou; 330.2 million was granted to Benin for the agricultural industrial complex of the SODAK organisation (Société Dahoméenne du Kenaf); 330 million was allocated to Togo for the improvement of its capital, Lomé; 64.9 million was received by Niger for the sinking of wells; 223.9 million went to ICODA (Industries Contonnières du Dahomey) for the construction of a textile factory at Contonou; Upper Volta received another 75.8 million for the acquisition of material for public works, and 127.2 million was allocated to the Public Building Society of the Entente for the construction of the House of the Council sited at Lomé.

The Intervention Budget of the Fund, unlike its loan scheme, does not guarantee loans *per se*, but it tries, within the resources at its disposal, to bail out members who, having started a project, are unable for financial reasons to complete it, either with subsidies, by underwriting all or part of the cost of projects, or by granting loans on concessional terms. It can also intervene to finance projects of general interest to the Entente.[55] In practice, the intervention budget is biased in favour of economic feasibility and other studies of common interest to the grouping. This can be deduced from the projects it has funded over the years, which include studies of marble quarrying in Benin and meat production within the Entente; the possibilities of an exchange of manufactured goods among the member-states; a study of endemic diseases and an advance payment to the state of Niger for its drilling programme. This is to mention only a few.[56]

It must be clear from the foregoing that the main advantage of the Intervention Budget lies in the fact that it supports projects of a social infrastructural nature which are not particularly attractive to private enterprise. For example, the study on the exchange of manufactured products within the Entente[57] is of interest to all the

54. See *Bulletin de l'Afrique Noire*, no. 586, 11 February 1970. Also see Thompson, *op. cit.*, 270.
56. For further information, see *ibid*.
57. *Bulletin de l'Afrique Noire*, no. 591, 18 March 1970. This issue provides an impressive summary of the study.

member-states but it was unlikely to have been undertaken by a profit-maximising private enterprise. The study assessed the tempo and areas of industrial exchanges within the grouping and identifies the industries likely to benefit from expansion on a regional scale. It notes that in 1967 oil products represented 12.4 per cent of intra-zonal exchange of industrial products while mechanical and metal processing accounted for 11.2 per cent. It classifies, *inter alia*, vehicles, aluminium sheets, cigarettes, beer and cement products as possible areas for the promotion and harmonisation of industrial exchanges. Thus, seen from the standpoint of its activities, the Mutual Aid and Guarantee Fund of the Entente can be said to be having some effect within the grouping, but whether this would materially and substantially influence the rapid development of the poorer economies within the Entente remains to be seen.

A key initiative which the Entente has taken towards the strengthening of economic co-operation among its members was the creation of an Economic Community of Cattle and Meat (ECCM) in May 1971, by an ordinance signed by the Presidents of the Entente states in Ouagadougou. The goal of the ECCM is 'to further together within a regional framework the commercialisation of cattle and meat within the boundaries of each country, and between the member-states and third countries, whether they be neighbours or not and especially those of the OCAM group'.[58] Naturally, the attainment of this goal falls squarely on the two bodies that implement the ECCM. First, there is a supreme policy-making body comprising the council of ministers, which has the responsibility not only of defining policy but of fixing the contributions of member-states. It convenes at least once a year and its decisions are taken by a unanimous vote of members. Below the Council of Ministers is an executive secretariat at Ouagadougou, whose responsibility is limited to the execution of studies and to the elaboration of programmes and propositions decided upon by the supreme decision-making body.

It is still early to comment on the performance of the ECCM. Indeed, there are still important questions to resolve before it becomes fully operational, such as those relating to customs and fiscal regulations. Despite several conferences the creation of the ECCM dragged considerably before the agreement was signed. There was the usual conflict between the interior and coastal states, which remained squarely opposed on matters such as the location of slaughter houses, fiscal and customs policy, prices for producers and to consumers, and investment priorities. The situation was further

58. *Ibid.*

complicated when the United States authorised its first large-scale loan of US $6 million to the Entente in early 1971 to support investment in livestock within the framework of the ECCM.[59] With the availability of this substantial sum, existing differences among the members over the preparation of protocols and selection of investment priorities gathered force. In the end the ECCM agreement of 1971 was signed, with such vital issues as customs and fiscal regulations still unresolved. Finally, it became clear that the United States loan would after all not materialise unless some structure was set up to justify it.

In these circumstances, it can hardly be said that the ECCM got off to a good start. Although each country had a long list of projects for which it was soliciting loans, these tended to concern the country's narrow internal interests. Petty economic nationalism must yield place to effective co-ordinated planning within the regional framework if the cherished goal of the ECCM is to be achieved.

5(c). *The Entente's political dimension; inter-state relations.* In reviewing the past activities and future aspirations of the Entente, one should not lose sight of the political dimension of the issues involved. As we noted earlier, the base of the Entente is more political than economic,[60] hence the dominance of political factors in its affairs. In general, the Entente states display some degree of political cohesion. Their regimes are moderate, favouring liberal economic policy and tending to present a co-ordinated front on African and world issues; they share the same colonial past; they all belong to the franc zone and, except for the Ghanaian 'enclave', form a geographically contiguous 'bloc'.

However, the degree of political cohesion existing between the Entente member-states is more apparent than real. For one thing, the conflict between the north and south, or the Sahel and coastal

---

59. The US aid to the individual Entente states (Benin, Ivory Coast, Niger, Togo and Upper Volta) between 1960 and 1968 amounted to $59.9 million but only $5.7m. of this was classified as regional aid, the rest being bilateral commitments. But since 1969 there has been a sharp policy shift from bilateral to regional aid, if only to minimise the political factors and to improve the economic effectiveness of aid. Total regional aid to the Entente from the United States stood at $6.6m. in 1969; $4.6m. in 1970; $13.8m. in 1971 and $20.0m. in 1972. See *Bulletin de l'Afrique Noire*, no. 591, 18 March 1970.

60. The 1961 agreement between France and the Entente states, which is one of the pillars on which the grouping stands, covered a range of issues, including defence, legal, social and economic matters. Upper Volta did not sign that aspect of the package deal which related to defence (see *Africa South of the Sahara*, 1971, London: Europa Publications, 89 and 119).

zones, has always existed. In more recent years, especially since 1970, a succession of crises has thwarted whatever political cohesion existed between the member-states of Entente. In November 1970, the Ivorian government felt compelled — ostensibly for security reasons in the wake of the Guinea invasion and students' demonstrations — to expel some 500 students[61] belonging to the Entente states, an act which did not endear President Houphouet to his Entente partners. Instead of a thaw, a further rift in inter-state relations developed in 1971, when Ivory Coast clearly failed to gain the support of its fellow Entente members for its policy of 'Dialogue with South Africa' at the OAU summit in Addis Ababa. Again, at the 1972 OAU summit, only Niger voted for the Voltaic candidate, Mr Malick.

This political disharmony must be seen against the background of the personality of the Ivorian President within the Entente. As indicated already, he is the hub on which the activities of the grouping revolve, and these activities merely seem to reflect his personal policies. This dominance had been resented in the past, but had not diminished. Indeed, the chief weakness of the Entente lies not so much in the inequalities of its partnership, the artificiality of its boundaries or the lack of group goals and organisational structure as in its overdependence on one man whose principal objectives are personal and national aggrandisement.[62] Given that Houphouet does not operate according to any fixed ideological guidelines, if he should suddenly decide that the game of maintaining the Entente is no longer worth the candle, or if the prosperity of Ivory Coast should falter and its subsidies cease, the organisation would doubtless dissolve. In more recent times, however, the Entente has demonstrated a measure of independence. Under a programme for the development of African enterprises being carried out through the development banks of the Entente countries and national promotion centres, foreign assistance has been received from several sources including a loan of $7.5 million from the United States Agency for International Development (USAID) and a transport study project financed by the Canadian International Development Agency (CIDA). By the end of 1974, 142 projects had been assisted under the programme.[63]

But the most immediate danger for the Entente's solidarity is the resurgence of Nigeria and its increasing role and influence in West

61. See T. Golan in *West Africa*, 30 October 1972.
62. See V. Thompson, *op. cit.*, 284.
63. UNCTAD, *Economic Co-operation and Integration Among Developing Countries*, vol. 2, 20 May 1976.

Africa and, to a lesser degree, Ghana's efforts to revive its economy and to repair the damage done to its neighbours by the expulsion of thousands of their nationals. If the Ghana government succeeds in reasserting its attraction for Upper Volta and Togo, and if Niger and Benin are drawn into Nigeria's orbit,[64] Ivory Coast may find itself relatively isolated in the region. However, whether this will happen or not depends largely on the existence of a more balanced intra-Entente development, the creation of the right political atmosphere within the grouping, and above all the state of inter-state relations in the wider context of West Africa following the formation of the ECOWAS.

What emerges from the preceding discussion is that the Entente Council has operated as a loose grouping in the face of difficulties. It could have some impact on the economic development of the area but only to a very limited extent. To start with, only in the case of Upper Volta does intra-area trade account for a large part (actually two-thirds) of its total exports. Nowhere else does it account for more than 10 per cent of trade; and in Ivory Coast it contributes only 3 per cent of total exports.

However, it would appear that, irrespective of the limited contributions made by a grouping of the Entente variety to the rate of economic growth, it is important to diffuse development in the area and to create the right political atmosphere and healthy inter-state relations that would make it possible for whatever potential exists to be fully exploited.

6. *The Organisation of Senegal River States (OERS)*

6(*a*). *Structure and Organisation.* The short-lived Organisation of Senegal River States (OERS), whose founder-members were Guinea, Mali, Mauritania and Senegal, was established by Convention in 1968. Its objectives were the maintenance of co-operative and peaceful relations among member-states, and the furtherance of economic development by co-ordinated planning and increased mobility of goods and people. It was intended to be a step towards the eventual creation of a wider regional grouping of West African states. The OERS hoped to realise its objectives by a common approach towards trade, fiscal and monetary policies, by a harmonisation of educational and training systems, and by the

---

64. Nigeria has a railway link with Niger. Nigeria and Benin recently agreed to establish joint cement and sugar industries, both of which are to be sited in Benin (see *West Africa*, 20 January 1975).

conclusion of an agreement to assure the 'right of establishment'.[65] Until its dissolution in November 1971, the OERS functioned on three levels: the Conference of Heads of State and Government, the Council of Ministers, and the organs of the Secretariat.[66]

At the highest level, the heads of each member-state met in ordinary session once a year to make general policy decisions and to examine recommendations made to them by the Council of Ministers; they could also meet in extraordinary session as often as necessary. Each, of course, had one vote and all resolutions had to be passed unanimously. The Conference established and adopted its own by-laws, and approved those of the other bodies of the Organisation.

The Council of Ministers, operating at the intermediate level, was composed of three ministers or plenipotentiaries from each member-state, and normally met twice a year, but could also meet by request in extraordinary session. Essentially, it was an institution of conception, execution and control. It prepared and proposed policy measures, and was responsible to the Heads of State. It too made decisions only by the unanimous vote.

Day-to-day management of the OERS was by an executive secretariat and three general secretariats. These offices were in Dakar, and were directed by the Executive Secretary who co-ordinated the operations of the three Secretaries-General and oversaw the Organisation's daily activities.[67] The Secretary-General for the Development of the Senegal River Basin was responsible for the promotion and co-ordination of studies and programmes for improvement, in conformity with the International Conventions of 26 July 1953 and 6 February 1964 relating to the development of the Senegal River Basin.[68] The Secretary-General for Planning and Development was responsible for preparing, presenting and executing harmonised and co-ordinated plans for the economic development and integration of the member-states. The Secretary-General for Educational, Social and Cultural Affairs was charged with parallel responsibilities. The Executive Secretary and the Secretaries-General were appointed for terms of three years by the Council of Ministers to whom they were responsible. Although not obligatory under the Convention, it was the practice that each

---

65. Statut de l'Organisation des Etats Riverains du Sénégal: amende (Dakar, 1970), articles 1–5. Hereafter, this document is cited as 'the Convention'. Also see R. Bornstein, 'The Organisation of Senegal River States', *The Journal of Modern African Studies*, 102 (1972), 267–83.
66. *Africa Research Bulletin*, 2203B.
67. See R. Bornstein, *op. cit.*, 268.
68. *Ibid.*

The Organisation of Senegal River States (OERS) 103

should come from a different member-state.

The OERS budget was prepared by the Executive Secretary and the Secretaries-General, but was adopted by the Council of Ministers. Financial contributions were determined by the Conference of Heads of State, on proposal by the Council of Ministers.[69] The organisation had an Advisory Committee composed of deputies from the National Assemblies of the four member-states, and representatives of various social and economic associations. But its functions were merely ancillary, and the committee had no actual power within the administrative hierarchy.

6(b). *History and activities of the OERS.* The history of the OERS was short and chequered. Although it existed for three years — 1968–71 — only in 1970 was there a period of intense co-operation. As we shall see below, the objective of bringing about co-ordinated economic development by sub-regional co-operation foundered in the face of political odds. The economic activities of the OERS were limited almost exclusively to the execution of feasibility studies for long-term development. In the middle of 1971 there were no major substantive economic projects in operation.

Consequently, substantive economic projects remained at the feasibility-study stages and were never implemented. Some of the more important planned projects will be briefly discussed here for the sake of interest.

Development of the Senegal River Basin. Because of their geographical position, the former OERS members have always been faced with great agricultural problems; none of them can produce enough food for the peoples within their own national borders. Thus the construction of a dam at Manantali in Mali (and possibly, another at the delta of the Senegal River) to irrigate the arid farmlands of the member-countries was considered an important step towards the solution of the area's agricultural problems. By March 1971, the result of studies commissioned by the OERS and executed primarily by UN experts covering all aspects of the project had shown that the project would cost a minimum of $100 million, and take at least six years to complete from the start of operations.[70] However, by the

---

69. The 1970/1 budget totalled CFA Fr.76 million (US$276,000) (see Bornstein, 269).

70. *Rapport de la 'Table ronde' sur les perspectives de développement intègre du bassin du fleuve Sénégal* (Dakar, 1971), 3. For details see R. Bornstein, *op. cit.*, 273–5. Apart from the provision of controlled irrigation, the project, if it were implemented, would contribute to transport improvement as well as provide additional power resources for industrial development in the area.

time the OERS ceased to exist no single organisation or country had expressed serious interest in financing the dam.

Monetary policy. As discussed earlier, Mauritania and Senegal are both members of the UMOA and the BCEAO, while Mali has a separate arrangement with France which has enabled it to regain a position in the franc zone. But the position of Guinea is quite different. It issues its own national currency which is non-convertible. Although nominally pegged at the value of the CFA franc before the 1969 devaluation, the Guinean franc is considered one of the weakest currencies in Africa. Thus, during the life of the OERS, not all the currencies of its members were freely convertible into each other.

In recognition of this problem for trade and payments, it had been hoped that a system of inter-state payments within OERS would be worked out. Article 3 of the OERS Convention clearly states: 'The Governments of the member-states of the OERS pledge, in the absence of a common monetary zone and free convertibility of their currencies, to facilitate inter-state payments in order to develop trade among member-states.' There can be no doubt that Mali and, particularly, Guinea would have greatly benefited from the implementation of such a pledge which, however, eluded the organisation. Given the instability of the Malian and Guinean francs and the large French influence in CFA monetary policy, the establishment of a monetary union or zone of free convertibility within the OERS proved a very difficult task. Besides, the level of inter-OERS trade (see Table V.6) is very low; hence the effect of such a monetary arrangement, at least in the short run, would be exceedingly marginal. Perhaps things might have changed in the long run, but time was not on the side of the OERS. Eventually, the organisation made no substantial progress in the field of monetary integration.

Industrial development. As already noted, it was the goal of all the OERS countries to achieve greater industrial development, and to this end the Ministers of Planning and Industry, in order to overcome the narrow base for growth within each country, decided in June 1970 to co-ordinate their plans in the industrial sector. Since they wished also to avoid unnecessary competition from similar products, they adopted a resolution calling for integrated and complementary development on the planning basis (*source*; Bornstein):

Guinea: Paper, tyres, aluminum, chemical electrolysis from marine salts

Mali: Metalworks, nitrate, sugar refinery, flour mill, alumina.

Table V.6
MALI AND SENEGAL: TRADE STATISTICS, 1967–69[1]
(US $000)

	1967	1968	1969
	Mali: exports		
Total	8,248	10,733	17,310
Guinea	14	—	85
Mauritania	96	108	76
Senegal	2,641	1,653	823
Total OERS	2,751		984
	Mali: imports		
Total	25,853	34,298	38,912
Guinea	17	25	55
Mauritania	92	122	153
Senegal	1,052	2,466	3,262
Total OERS	1,161	2,613	3,470
	Senegal: exports		
Total	137,286	151,384	123,696
Mali	92	1,908	4,066
Mauritania	1,450	1,674	1,866
Guinea	236	216	516
Total OERS	1,778	3,798	6,448
	Senegal: imports		
Total	157,558	180,990	198,666
Mali	692	384	617
Mauritania	34	118	33
Guinea	4	151	179
Total OERS	720	653	829

1. Complete statistics are not readily available for Guinea and Mauritania, but their OERS trade figures may to some extent be gathered from this table.

Source: Statistical Office of the European Community, *Associated Foreign Trade Yearbook*, Brussels, 1967–9, cited in R. Bornstein, *op. cit.*

Mauritania: Metalworks, copper, cement, plasterworks.
Senegal: Petrochemicals, pharmaceuticals, chemical electrolysis from marine salts, polymers.

The Secretary General for Planning and Development called for the 'complete denationalisation of these industries, whatever their geographical locations', and further suggested that measures be taken to provide for free circulation and consumption priority within the sub-region for goods produced by these industries; effective protection of these products from foreign competitors; reinvestment of profits in OERS-integrated industrial projects and their distribution in equal shares to the member-states; and common management of these industries by responsible nationals appointed by the Council of Ministers.[71]

Attempts were made to translate these proposals into action. In December 1970 the OERS concluded an agreement with the UN Industrial Development Organisation (UNIDO) to promote and accelerate integrated economic development, with special reference to the complementary proposals (above) made by the four Ministers of Planning and Industry.[72] But integrated intra-zonal development needed time to be accomplished. The political differences of the OERS members, reinforced by the diversity of national economic policies and problems, acted as a brake on any meaningful progress towards co-ordinated intra-area industrial development.

Livestock and animal programmes. The OERS countries possessed great potential for the development of meat-processing and tanning industries (see Table V.7). In 1968, an Inter-State Commission for Livestock and Animal production was formed to co-ordinate the efforts of the four countries in evolving more efficient management, and among the questions studied were breeding techniques, feeding systems, marketing methods, disease prevention, applied research and personnel training.[73] The Council of Ministers later approved a recommendation for common legislation to combat animal disease which, had it been enacted by the National Assemblies, would have been the first co-ordinated effort to enact harmonised legislation for the grouping. Before the disintegration of the OERS some measure of success has been achieved in the livestock sector. A bilateral Mali-USAID poultry project was expanded to the sub-regional level. Similar projects involving the construction of hatcheries and the

---

71. Bornstein, *ibid.*, 279.
72. However, this did not commit UNIDO to specific financial support and technical assistance outside its normal practices and procedures (see Bornstein).
73. *Ibid.*, 280. Also see *Réunion de la Commission inter-état de l'élevage et de production animals* (Conakry, 1970).

Table V.7
OERS MEMBER STATES: ESTIMATED ANIMAL POPULATION
1969-70

	Bovine	Ovine	Equine	Asinine	Camel
Guinea	1,450,256	605,559	51	759	—
Mali	4,063,900	7,813,900	603,290	203,011	77,700
Mauritania	1,800,000	6,000,000	15,000	150,000	700,000
Senegal	2,525,000	2,520,000	191,000	172,000	8,000

Source: R. Bornstein, op. cit.

introduction of modern breeding techniques were being built and were expected to be operational by the end of 1971.

Aside from the sectors discussed above, the OERS Secretariat developed numerous other proposals for co-operation in the educational, cultural and social fields. But, like others, these projects were very ambitious; and there was neither time nor a conducive spirit of harmony to enable them to materialise.

6(c). *Political dimension and inter-state relations.* The familiar but debatable maxim that political co-operation must precede regional economic development seems to have been vindicated in the precarious existence of the OERS. Its short life was punctuated by recurrent political disputes which eventually sounded its death knell.

Shortly after the Convention was signed at Labé in Guinea on 24 March 1968, the Organisation was confronted by its first political crisis. On 19 November 1968 the socialist government of Mali and its President, Modibo Keita, were overthrown in a military *coup d'état* which installed Lieutenant Moussa Traoré as the new head of state. Although military coups had become a common feature in post-independence Africa, this was the first coup experienced by the four countries under discussion. It therefore had profound psychological and political effects on the other OERS member-governments. The fall of the Keita regime was of special concern to Sekou Touré of Guinea, who had lost another close socialist ally when Kwame Nkrumah of Ghana was deposed in 1966. Having now lost his most trusted comrade and finding himself isolated, Sekou Touré tacitly withdrew his support for the OERS, substituting rhetoric for positive action. The Guinean policy of non-co-operation effectively paralysed the new organisation for the rest of 1968 and all of 1969. Little, if anything, of practical importance, was achieved by the OERS during this period, essentially because under the Convention, as noted above, all major policy decisions had to be made with the unanimous vote of all the four heads of state — but all the four never met between November 1968 and the end of 1969.[74]

The second phase in the OERS inter-member relations was marked by a determined effort at resuscitation. Through a vigorous diplomatic campaign by other members, spearheaded by Mauritanian President Ould Daddah, the four Heads of State met in the Guinean capital of Conakry on 3 February 1970, and the Convention was reaffirmed and amended.[75] It was thought that after the conference members would forget their political differences and that the organisation would enter a new era of harmony and progress. Indeed, there followed a proliferation of specialised meetings to draw up OERS programmes in a wide range of fields (see above). All four countries regularly reaffirmed their commitment to the organisation in spite of disguised political bickering (this time between Mauritania and Senegal). However, this spell of intense activity did not last long. The invasion of Conakry on 22 November 1970 by a small outside force under Portuguese officers[76] signalled its end. Although the Council of Ministers, which met in extraordinary session a few days after the attack, unanimously deplored the 'barbarous aggression', the solidarity which the incident generated quickly proved short-lived.

The invasion and its aftermath at one and the same time halted the phase of intense co-operation, precipitated the phase of disintegration. Several weeks after the invasion, Guinea accused Senegal of massing troops for the purpose of launching a new attack. This was strongly denied by Senghor's government but the allegation created a serious political rift between the two countries. Sekou Touré did not attend the Conference of Heads of State held in Bamako, Mali, on 18 January 1971. Not surprisingly, the Conference was a failure since no decision could be taken on the numerous projects drawn up in 1970 and no plan could be charted for the future. Until Senegal resigned from the OERS on 30 November 1971, presumably out of frustration, no durable détente existed between the two members.

Realising that the OERS as then constituted was no longer an effective body, a Ministerial Conference of the Organisation was called, which took place at Nouakchott, Mauritania, on 29 November 1971, again without Guinean representatives being

---

74. *African Research Bulletin*, 1583C.
75. *Ibid*. This was the first meeting of all four heads of state since the signing of the convention in 1968.
76. A UN investigation has confirmed that this force originated from the neighbouring Portuguese colony of Guinea-Bissau. Although the invading force, which was beaten back, included some Guinean exiles, the UN has blamed Portugal for the ill-fated invasion (See *Africa south of the Sahara, op. cit.*, 373).

present. It was at this two-day conference that Senegal resigned. The remnants merely formalised the death of the organisation when they satisfied themselves 'that the OERS no longer answers to the needs it was founded to satisfy'.[77] In March 1972 the Heads of State of Mali, Mauritania and Senegal formally signed an agreement winding up the OERS.[78]

6(*d*). *A new organisation created: OMVS (Organisation for the Development of the Senegal River).* To salvage whatever could be salvaged from the defunct OERS, a new Organisation for the Development of the Senegal River (OMVS) was established in Nouakchott on 11 March 1972 after discussions by the Heads of State of the founder-members (Mali, Mauritania and Senegal). The Organisation was declared 'open to all states through which the river flows provided they accept the spirit and letter of the Convention'.[79]

The structures of the OMVS are lighter than those of the OERS. The General Secretariat is the key structure but it is capped by a ministerial council which has more of a technical than political influence, and inherits most of the powers of the former Heads of State Conference, which had to meet obligatorily each year. In future, the member heads of state will only meet when it is felt necessary. Because of its strong desire to ensure that it will be effective, the new body has a strong legal foundation since the new convention determining the statute of the Senegal river (which has been declared an international waterway) and its tributaries, cannot be denounced by its trustees for the next ten years.[80] The Secretariat of the OMVS, headed by a Secretary-General, is in Dakar.

In many respects, the orientation and structures of the new organisation are more modest and less political than was the case with its predecessor. Having learnt from their past experience, the OMVS founders opted for a less ambitious goal. Though calling for the co-ordination of its members activities in almost every possible field, the OMVS 'only fixes one definite economic objective: the planning and development of the Senegal river'.[81]

To realise this goal, a private firm was hired to make a synthesis of all the studies so far carried out on the development of the Senegal river basin. The consolidated report, which was approved in May

77. *Ibid.*, 2293.
78. *Ibid.*
79. It is doubtful whether Guinea would join the new organisation since Sekou Touré has declared his preference for security as against foreign currency after the alleged plan by Senegal to attack Guinea (*ibid.*).
80. *Ibid.*
81. *Ibid.*

1974 by the Ministerial Council, incorporates a development programme extending up to the year 2011 and will require financing of CFA Fr.800 billion.[82] The first stage of the programme is to be completed by 1982 and will represent an investment of CFA Fr.42 billion. This stage involves the construction of six major works, including the Manantali dam in Mali on the Bafing river, one of the sources of the Senegal, which is intended to regulate the river's flow and to produce hydro-electric power for the development of the iron, copper, phosphate and bauxite deposits in the upper part of the basin. The programme for the development of the Senegal River up to the year 2011 contemplates the gradual addition to the area under cultivation of 428,000 hectares (45,000 in Mali, 140,600 in Mauritania and 428,400 in Senegal). The crops to be grown on these lands will be mainly rice, wheat, sorghum, maize, niebes, sugar cane, vegetables and fodder. This latter product will permit a considerable increase in the number of livestock and it will thus be possible to satisfy the meat requirements of the three countries and to cover a substantial part of their milk supplies. It is also planned to construct 232 plants for processing agricultural products and to develop the mineral resources of the Senegal River Basin.

However, in view of the interdependence of these projects and the possibility that the exercise by a member-state of its sovereign rights over an essential project might effect the proper functioning of the system, the Council of Ministers agreed that the projects of common interest would be the common property of the Contracting Parties.[83] But the key problem facing the programme is the question of financing, and unless this improves, some of the projects will not be implemented.

The OERS, whatever its aims, achieved very little. It ought to have done better and could have done so, but the framework of political and ideological understanding between its member-states was too tenuous to bear the strains of international co-operation. In the end, it died a victim of politics rather than of economics. Its failure is a further lesson for the future of economic integration in West Africa. Although a new organisation has been set up in its place, and special arrangements have been made to ensure the success of the OMVS, no firm guarantees can be offered at this stage regarding its success.[84] Ultimate success must depend on workable machinery

---

82. See UNCTAD, *Economic Co-operation and Integration among Developing Countries*, vol. 11, May 1976 (TD/B609/11), 49.
83. *Ibid*.
84. Indeed, secret talks have been going on to defuse a potentially dangerous crisis

for co-operation, and, more important, in the existence of solid political detente among members of the grouping.

7. The Senegambian case

7(*a*). *The rationale of economic integration between Senegal and the Gambia.* So far we have been discussing various forms of economic co-operation between the countries which were formerly parts of French West Africa. This section briefly discusses one of the earliest attempts at bridging the co-operation gap between the Anglophone and Francophone states of West Africa.

The anomalous geographical position of the Gambia reflects colonial policy in West Africa which we have discussed earlier.[85] Completely surrounded by Senegal except on its seaward margin, the Gambia forms an irrational intrusion into the much larger country of Senegal and largely isolates the southern region of Casamance from the rest of Senegal. Because of the Gambia's strange geographical position, some form of integration with Senegal has often been suggested as its natural destiny.[86]

Following Senegalese independence in 1960 and the prospect of early independence for the Gambia, the urge for a closer relationship between the two countries gathered force. The case for the Senegambian integration centred on three main considerations.[87] First, there were doubts about the economic viability of the Gambia on its own. This fear is by no means unjustified for, as the summary data in Chapter II show, Gambia is a tiny and poor country by any standards. At present Britain provides more than 60 per cent of its development budget finance, and this financial dependence for

---

threatening to damage the OMVS. Senegal and Mauritania are both claiming Todd island, a small strip of land in the Senegal river on Senegal's northern border. Except for its lucrative cattle trade, the island is of little importance but the future of the OMVS may well depend on how the crisis is resolved (*West Africa*, 9 June 1975).

85. See Chapter 1. Ethnically the people of the Gambia and Senegal are the same. Indeed, for a brief period, between 1765 and 1783, much of today's Gambia and Senegal formed a single British colony of Senegambia. That they are now two separate countries is explicable mainly in the context of rivalries between the European colonisers (see also P. Robson in Hazlewood (ed.), *op. cit.*, 115).

86. Even during the colonial period several abortive British and French proposals, which envisaged the exchange of some other French colony for Gambia so as to permit its incorporation into Senegal, were put forward. But they were all doomed to failure due to differences between the two colonial powers (see J.D. Hargreaves, *op. cit.*, 145–95).

87. See Hazelwood (ed.), *op. cit.*, 115–28. This article provides a well-informed discussion on the 'Problems of Integration between Senegal and Gambia' and this section (V.7) draws on it. Also see A. Hughes, 'Senegambia Revisited' in *Senegambia* (proceedings of a colloquium at the University of Aberdeen), 1974, 139–70.

development expenditure is likely to continue.[88]

Secondly, as an enclave within a country with twenty times its own area, the Gambia is an inconvenience to Senegal, both because of the physical division it creates and because it is a smuggling base. It was estimated around 1960 that smuggling through the Gambia into Senegal accounted for something of the order of 25 per cent of the Gambia's total exports of domestic produce.[89] Although the amount of goods so smuggled is very small in relation to Senegal's imports — less that 1 per cent — the practice, which has become Gambia's economic life-blood, shows little signs of declining. In fact, far from being stamped out, smuggling is assuming a new dimension. 'Transistors go to Senegal, groundnuts come to Gambia, both ways the Gambia makes money' — because Gambians enjoy freer access than the Senegalese to the world market. As a Senegalese Finance Minister, Jean Collin, once remarked: 'If the Gambians had all the transistors they imported, there would be 212 transistors for every 1,000 Gambians compared with 5 for every 1,000 Senegalese.'[90] Consequently, illegal border trade has long been a constant, if not a permanent, source of irritation and friction between the two countries.

The present economic frontiers are also disadvantageous to both countries in other respects. For Senegal they mean a partial isolation of the southern province of Casamance and inability to use the Gambia river fully. On the other hand, it is argued that the Gambia cannot exploit its main natural asset — the river basin — and that Banjul (known until April 1973 as Bathurst) is deprived of the opportunity to serve as the port for a large economic area to which it is properly suited. So, in short, it is held that adequate use of the economic resources of the Gambia and Senegal requires close cooperation between them, if only to overcome the artificiality of the existing political and economic frontiers.

The third factor is political. There has been some fear in Senegal that the Gambia could become a staging base for the operations of banned political parties or for subversion from outside. For her part,

---

88. Of the £5 million required for the 1967–71 development programme £3.2 million was provided as an interest-free British loan, and over this period 90 per cent of development expenditure had been covered from British sources (see *Africa South of the Sahara, op. cit.*, 333). Britain was also committed to meeting some 50 per cent of the outgoings of the Third Development Programme, 1971/2 to 1973/4. But as we shall show later, British aid has been redirected away from the recurrent budget into the Development Budget (A. Hughes, *op. cit.*).

89. N.G. Plessz, *Problems and Prospects of Economic Integration in West Africa*, Montreal: McGill University Press, 1968, 62.

90. Jeune Afrique, *Africa*, 1971 *op. cit.*, 259.

the Gambia, lacking an effective army, recognises its extreme vulnerability from a military point of view — in the event of an attack.

It is against this sombre background of fears and mutual suspicion that talks on the possibility of closer economic co-operation has often been held since the early 1960s. An Inter-Ministerial Committee was set up in 1961 by the two countries for mutual discussion of matters of joint interest. This Committee, is still maintained, but has not achieved a genuine thaw in the field of economic integration between the two countries. Nonetheless, it initiated discussions on the subject which invariably led to the commissioning of a report from the United Nations to consider the various possibilities of association between the two states.

7(b) *The United Nations report.* During the later part of 1963, a UN Technical Assistance Mission was sent to the Gambia and Senegal at the request of their respective governments to investigate the possibilities of a closer association between them. The UN report,[91] which was submitted early in 1964, dealt with political, economic and fiscal aspects of association, and was followed by a supplementary report by the FAO on co-ordinated agricultural development in the Gambia river basin.

On the political front, the UN report suggested three alternative forms which association between the Gambia and Senegal might take. The first was full integration of the Gambia as the eighth Senegalese or Senegambian province. But this option was ruled out as unacceptable to the Gambia and not to be entertained 'until a long period of friendly and fruitful collaboration between the two countries has elapsed'. The second alternative envisaged the formation of a loose Senegambian federation. The report recommended a federal government with powers for the initial period limited to defence and overseas representation and with complete autonomy in other respects for the federated states. Progress after the initial period would depend on the wishes of both states. The authors of the report evidently favoured this alternative but had reservations as to whether it would prove acceptable to the Gambia. Assuming that these two alternatives would be considered premature and/or unacceptable, the report advocated the establishment of a Senegambian entente — possibly in the form of a treaty relationship — which would involve neither the creation of a new

---

91. UN, Department of Economics & Social Affairs, *Report on the Alternatives of Association between the Gambia and Senegal*, March 1964. For commentaries on the report, see N.G. Plessz, *op. cit.*, 120-8.

state nor the impairment of the sovereignty of each existing state, and would be a practical step and prelude to eventual closer integration.

In the economic sphere, which is of particular interest to us here, total economic integration was not seriously contemplated in the report since this would have required a considerable degree of political unity which had been ruled out as not being feasible. Therefore, the form of economic integration advocated in the report was customs union, embracing fiscal harmonisation, and ultimately monetary integration, with the Gambia making the adjustments. But the immediate practical problems of a sudden economic integration, such as the administrative difficulties arising from the introduction of the complicated Senegalese regulatory system into the Gambia overnight and the effects of such changes on the cost of living, were clearly recognised. This led the UN report to endorse a gradual economic association of the two countries beginning in areas where agreement was feasible and easy to attain, and gradually building up to a more advanced form of association. It was thought that a developing economic association would help to promote a gradual rapprochement in the political sphere as well.

What eventually emerged as the practicable form of economic co-operation between the two countries in the transitional phase was a free-trade area with import restrictions in the Gambia. Under this customs union frontiers would be abolished and the Gambia given an overall import quota, based on recent import levels, to which reduced rates of duty would apply, corresponding initially to the rates hitherto levied. Provisions would also be made to curb smuggling. On monetary matters, the report took the view that while the currencies of Senegal and Gambia ultimately would have to be unified, this was not considered urgent, partly because of Senegal's membership of UMOA, which removed its autonomy in the monetary field, and partly because the Gambia belonged at the time to the now defunct West African Currency Board. Even so, the report did not deem the long-term technical problems of monetary unification insuperable.

As we shall soon see, these proposals failed to satisfy the governments of Senegal and the Gambia, and very little has come of them.

7(c). *The benefits of economic integration in Senegambia.* In theory, there are many economic gains to be derived from the integration of Senegal and Gambia. To begin with, the Gambia river could be used to transport the Senegal groundnut crop down to the coast for export, although the use of the river does not necessarily require the vertical integration of the economies of the two countries. Many other African countries, especially the landlocked ones, use the transport system of their coastal neighbours without formal tariff unification.

Another important case for closer association concerns the integrated development of the Gambia river basin in relation to irrigation. The FAO report, mentioned above, discussed the benefits that would accrue from the construction of a storage dam in the upper catchment area of the Gambia river, particularly to provide irrigation facilities and a power supply. Again, there is no *prima facie* case for assuming that the development of the Gambia water resources can only be undertaken within the framework of a customs union: it could equally be undertaken by an inter-governmental agency, and indeed experience in the region bears eloquent testimony to this view. The Inter-State Committee for the Senegal River Basin, founded in 1963 and later replaced by the OERS — all discussed already — had similar functions.

However, the more important consideration for the Gambia river projects relates to their economic feasibility. Technically they may appear quite practicable but, given the limited market base in the area and the difficulty of financing such projects, their economic feasibility may be called into question. Thus it appears that the FAO report, which proposed these projects, assumed that a customs union would be a good thing for all. But, if experience is any guide, the degree of benefit that a small country like the Gambia would derive from an integration scheme in terms of additional income will depend on the intra-union 'backwash and spillover effects'.[92] On balance, these can be unfavourable for a very small country joining a larger area — unless, of course, an adequate system of compensation is built into the integration agreement to reverse such a trend. For example, under the transitional arrangement proposed by the UN report, it would be necessary to introduce rationing in the Gambia to control prices, and this would in turn mean an increased administration cost, not to mention the revenue effects of tariff disarmament. These costs may not be easily offset by the gains accruing to a 'junior' partner in a *laissez-faire* integration scheme.

Commenting on the recommendations of the UN report, Professor Robson concludes that 'a transitional free-trade area as the prelude to a simple customs union offers Gambia no obvious advantages and some evident immediate disadvantages in the form of higher administrative costs. It would not be sensible to enter such an arrangement without some more equitable distribution of the

---

92. For a discussion on these effects see G. Myrdal, *Economic Theory and Underdeveloped Regions, op. cit.*, 23-49. See also A.O. Hirschman, *op. cit.*, 187-90.

direct costs and benefits of the change-over to the two countries ...'[93] He suggested, instead of the UN recommendation, the establishment of a full free-trade area which would permit each country to maintain its own tariff or alternatively the institution of a free-trade area in local agricultural produce only. The former alternative, unfortunately, fails to allow for the fight against smuggling, and Robson admits this, retorting that 'even the United Nations proposal involves accepting the continuance of smuggling for an indefinite period.'[94] This weakness notwithstanding, it is contended that either of the two alternative suggestions would offer a reasonable expectation of long-term advantage and no immediate disadvantages to both countries. While realising that a free-trade area may be regarded as a second best solution compared with a customs union, it is clearly underlined that realism demands no closer economic integration between the two countries at this point in time.

If one views the situation with detachment, it is difficult to escape this verdict — although the search for a more satisfactory arrangement continues as discussed below.

7(*d*). *Recent progress towards integration.* For more than seventeen years after the UN report, the associational relationship between Senegal and the Gambia made little progress, except in the sociocultural field. But, as we shall see later, the unexpected (abortive) coup in the Gambia in 1981 dramatically forced the two states to a hurried confederation.

The report was transmitted to the governments in March 1964, and talks were held between the two governments in Dakar in May of that year to examine the alternative proposals. At these meetings the Gambia came up with proposals on the political front, analogous to the third alternative in the UN report; these called for a confederal structure in which would be vested responsibility for defence, foreign relations and overseas representation. This was admittedly unacceptable to Senegal, which countered with proposals envisaging the eventual political integration of the Gambia with Senegal; but these in turn were unacceptable to the Gambia. In the impasse that ensued, the two countries settled on a treaty relationship covering foreign affairs and defence only.[95]

The defence agreement provides for mutual assistance in the face of any form of external threat, the establishment of a joint Senegal-Gambia Defence Committee with a permanent Secretariat, and

---

93. P. Robson, *Economic Integration in Africa*, London: Geo. Allen & Unwin, 1968, 285.   94. *Ibid.*, 284.   95. *Ibid.*, 126.

Senegalese assistance in training any Gambian military or paramilitary units. The foreign policy agreement provides for an exchange of resident ministers, representation of the Gambia by Senegal as and when directed by the former, and a Joint Committee on Foreign Affairs with a Secretariat meeting every three months to harmonise the approaches of the parties to all matters of importance in foreign affairs. While providing a useful framework for co-operation, these agreements do not in any significant sense impair the sovereignty of either country.

The Gambia's rejection of the recommendations for a transitional free-trade area with quotas seems to be based on the fear that integration might not create a sufficient increase in trade and economic activity in the Gambia to accommodate the consequent extra administration costs and loss of revenue through tariff cuts. For Senegal any form of economic association that would not lead ultimately to full integration of the Gambia seems out of the question. Thus the 1964 Agreements were not used as stepping-stones to closer relations; the two countries clearly dragged their feet over implementing even such modest acts of co-operation.[96] The foreign policy agreements, for instance, have never really worked: the joint committee has lapsed, there is no joint representation overseas, and the two states pursue separate and sometimes contradictory policies. Similarly, little has come of the defence agreements.

In accordance with the UN report's recommendations, the two governments signed, on 19 April 1967, a treaty of association defining the bodies responsible for promoting and extending co-ordination and co-operation between their two countries in all fields.[97] The treaty provides for the setting up of three bodies: a Conference of Heads of State, an inter-state Ministerial Committee and a permanent Senegambian Secretariat. The conference of Heads of State meets, in principle, once a year in Banjul and Dakar alternately, to define general guidelines and survey the state of co-operation between the two countries. Within the framework of the policy so defined, the inter-state Ministerial Committee's task is to study all measures to strengthen co-operation and solidarity between the two countries and to submit them for approval to the two governments. The Committee meets at least once a year. In practice, it has met twice a year, alternately in each country. The Secretariat, with its headquarters in Banjul, is a permanent investigation, liaison and information body entrusted with the implementation of the decisions of the Ministerial Committee to

---

96. See Hughes, *op. cit.*, 149–50.
97. See Sy, *op. cit.*, 128–9.

which it is responsible. Senegal has agreed to pay 75 per cent of the cost of the Secretariat out of its provisional budget of CFA Fr.31,360,000, with the rest coming from the Gambia. This was agreed at the second conference of the Committee on 29 January, 1968.[98]

When these institutions were set up, a series of agreements was concluded covering ever wider fields. The permanent Secretariat endeavoured to implement such agreements and still continues to do so but there have been difficulties in implementation. On 10 June 1967, the two governments signed a cultural agreement to develop as far as possible the relations between the two countries in the fields of university and school education, science, technical matters, sports and culture, in order to contribute to a better mutual understanding of their respective cultures and their activities in these fields.[99]

Since 1968 the Secretariat has drawn up an Implementation protocol to this end each year and certain cultural and social activities have been carved out within the framework of the Implementation Protocol to the cultural agreement. The Secretariat, working in close co-operation with those responsible for education, contributed to the setting up of an Advisory Co-ordinating Council for Youth and Sport. Its achievements have been modest, and include the building of a Senegalese school, containing 450 pupils more than 60 per cent of whom are Gambians; the organisation of sports; and the annual organisation of youth caravans in one or the other of the two countries.

Notwithstanding their importance, socio-cultural affairs represent only one aspect of the Senegambian co-operation. There is the more complex economic field, but here the only project worth mentioning is the development of the Gambia river basin. The purpose of this project, as indicated earlier is to make a hydrological assessment of the river with a view to developing its basin for hydroelectricity, shipping and irrigation. The study was initiated in 1970 with the aid of funds from the UN Development Programme (UNDP) and completed seven years later.[100] Although the role of the Secretariat in directing this project has been notable, it is still a far cry from the optimistic forecasts of massive dams providing unlimited power and vast acres of irrigated land.

The move towards closer co-operation between the two countries

---

98. C. Legum and Associate (eds), *Africa Contemporary Record*, 1968–69, London, 486 and 583.
99. The Secretariat itself was not set up until a year after it had been decided upon in the 1967 Treaty, and the Gambians do not seem to have been consulted over the selection of its Senegalese director (see Hughes, *op. cit.*, 150, and Sy, *op. cit.*, 129).
100. *Ibid.*

has been remarkably slow. Indeed, the associational relationship, which has characterised the position between them in the post-1964 period, has recently undergone stress. One important factor in this has been a certain disenchantment with integration on the Gambia's part, accompanied by a period of coolness and distance from its putative federal partner. This change of heart dates from around 1969 and has led to permanent changes in Gambian thinking on ultimate union with Senegal. The perennial issue of smuggling and the related border 'crisis' of 1969–71 confirmed Gambian fears that Senegal was bent on assuming a dominating role in the partnership, and although little hard evidence was produced to substantiate these fears, they have been sufficient to lead to divergences in Gambian foreign policy.

There is also the fact that from the Gambian point of view, the Senegambian idea has been more of a pragmatic response to a set of difficulties facing the colony-state in the early 1960s than a self-negating idealistic commitment to the goals of continental unity. The prime impetus was unquestionably a concern for the economic viability of the colony following the withdrawal of British rule, together with its grants-in-aid. The anxiety and uncertainty created by this situation turned Gambian eyes towards Dakar, since Senegal then seemed to offer a way out and Senegal for its part seemed anxious to form a close political and economic union with its tiny neighbour.

However in 1965, the year of Gambian independence, the groundnut crop — the mainstay of the country's economy — increased both in size and value, and despite later fluctuations, this steady growth has been maintained. Thus the Gambia was able to manage without a British grant-in-aid in 1966/7 and has had no recourse to it thereafter.[101] The emergence of this spirit of self-sustenance damped any desire for rapid and close union with Senegal. Furthermore, the Gambia, as we noted earlier, has justifiable reservations about the immediate benefits of economic integration with Senegal. Given Senegal's more developed and export-oriented industrial sector, Gambians feel that the Senegalese would require the abolition of their cheap import policy and the subsequent ending of the clandestine trade with Senegal.

The associational relationship between the two countries has assumed a new and dramatic turn following Kukli Samba Sanyang's violent attempt on 30 July 1981 to seize power in the Gambia and to set up a National Revolutionary Council. It was with Senegalese

---

101. Fear about the termination of British aid after independence turned out to be unfounded. It was merely re-directed into the Development Budget (see footnote 88).

120 The performance of existing integration schemes

political support and military intervention that Sir Dawda Jawara was restored to power in the Gambia — not surprisingly, at a price. The price is the unconditional acceptance of the idea of confederation, and indeed the Confederation Agreement between the Gambia and Senegal was signed in Dakar on 17 December 1981. Although this seems to be a realistic solution to the Gambian politico-economic problem, it is based on expediency rather than free choice. Thus there is a smouldering fear that, when the dust settles, the Gambians may wish to reassert their own political sovereignty.

8. Inter-state functional organisations

Aside from the more formal forms of economic integration discussed above, and the ECOWAS which is treated separately in the next chapter, there exist some inter-state functional organisations in West Africa (Table V.1). The more important ones, which are summarised here, are the River Niger Commission, the Lake Chad Basin Commission, the Mano River Union and the West African Clearing House. The functional organisations, unlike the other integration schemes, are generally specific and limited in their objectives, sometimes involving little or no integration in the field of fiscal, monetary or labour policies. Co-operation is often centred on specific issues or development project(s).

8(*a*). *The River Niger Commission.* The Niger, which is more than 4,000 km. long, forms a very extensive basin spreading over nine states, including, Benin, Cameroon, Chad, Guinea, the Ivory Coast, Mali, Niger, Nigeria, and Upper Volta. A conference of these states with the exception of Cameroon and Mali was held in Niamey on 15–16 February 1963. A United Nations study served as the basic document for discussion. At a second meeting held at Niamey on 24–26 October 1963, all the nine states including Cameroon and Mali signed an Act concerning navigation and economic co-operation between the states of the Niger Basin, known as the 'Act of Niamey'. This legal instrument abrogated the General Act of Berlin of 26 February 1885, the General Act and Declaration of Brussels of 2 July 1980 and the Convention of Saint-Germain-en-Laye of 10 September 1919, insofar as they concerned the River Niger, its tributaries and sub-tributaries.[102] For the contracting parties, these earlier agreements were no longer compatible with African interests and aspirations.

---

102. UNCTAD, *Economic Co-operation and Integration among Developing Countries, op. cit.*, 44–5.

The Commission is an inter-governmental organisation with the responsibility of promoting, encouraging and co-ordinating studies and programmes related to the development of the Niger basin. In other words, the Commission's functions are to maintain liaison between the member-states to ensure the most effective use of the waters and resources of the basin; to collect, evaluate and disseminate data on the basin; to examine projects prepared by the member-states; to recommend to governments plans for common studies for the judicious utilisation and development of the resources of the basin; and to draw up general regulations to ensure implementation of the principles set forth in the Act of Niamey.

The Commission has two organs: the Commission proper and the Administrative Secretariat. The former is composed of nine commissioners, one for each contracting party, and arrives at its decisions by a two-thirds majority. The Commission meets at least once a year. Working under it is an Executive Secretariat located in Niamey; the Executive Secretary holds office for a period of three years.

The achievements of the River Niger Commission so far have been very modest. American and Canadian assistance, *inter alia*, has helped to build a bridge over the Niger River linking Gaya in Niger with Malanville in Benin and to build a river port near Gaya. Togo enjoys an observer status with the Commission.

8(*b*). *The Lake Chad Basin Commission*. The Lake Chad Basin Commission came into legal existence on 22 May 1964 after the heads of state of Chad, Niger, Nigeria and Cameroon had signed the treaty establishing it in N'Djamena (formerly Fort-Lamy), Chad, which is also the headquarters. The Chad basin covers an area of about 550,000 square km.

The functions of the Commission are to collect, evaluate and disseminate information on proposals made by the Contracting Parties; to recommend plans for common projects and joint research programmes; to maintain liaison with member-states with a view to more effective use of the water; to draw up common rules for navigation, and to promote the settlement of disputes among members.

The two instruments of the Commission are the Commission itself and the Administrative Secretariat. The former consists of eight commissioners, two from each member, and meets twice a year. Decisions are taken unanimously. The Secretariat is headed by an Executive Secretary appointed for a three-year term by the Heads of State on the recommendation of the Commission. The Commission established a Development Fund to finance the planning, preparation and implementation of development projects, and the repayment of loans contracted by it. The resources of the fund are derived from an annual contribution from each member, amounting to not less than 30,000 Units of Account.

The Commission has four agricultural development centres, one in each member-country. Nigeria has spent about US$48.6 million (30 million naira) on its own Lake Chad Sprinkler Irrigation Scheme since it took off in 1976 in an attempt to irrigate some 50,000 hectares for the growth of wheat, maize, tomatoes, sugar cane and groundnuts.[103] The Commission has also sponsored a number of studies related to the development of agriculture, irrigation, fishing and transport in the region surrounding the lake. Similarly, implementation of a re-afforestation plan to combat the effects of the drought and the advance of the desert in the area was started in October 1974. The Commission is also establishing a telecommunications system which is intended to link N'Djamena, Fotokol, Fort Foureau and Maiduguri.

8(c). *The Mano River Union*. The history of the Mano River Union made up of Liberia and Sierra Leone goes back to 1967 when the two countries started to explore the possibility of establishing closer trade links and economic co-operation. The resulting bilateral negotiations gathered force in 1971 and 1972, when the governments of the two states requested the United Nations Development Programme to set up a joint UNCTAD, UNIDO and FAO mission to study the possibilities of co-operation, particularly in trade, industry and agriculture. The mission investigated the matter and recommended the establishment of a customs union between the two countries. The recommendations were considered and approved by the Joint Ministerial Committee of the Union at Monrovia in September 1973. On the basis of the Committee's recommendations, the heads of state of Liberia and Sierra Leone signed on 3 October 1973 at Malema the 'Mano River Declaration', whereby their respective countries undertook to establish a customs union,[104] and the accompanying protocols dealing with specific matters.

The objectives of the Mano River Union are to expand reciprocal trade through the elimination of existing barriers, to promote co-operation for the expansion of international trade, to create conditions favourable to an expansion of the productive capacity of the area, including the progressive development of a common protection policy and co-operation in the creation of new productive

---

103. See *New Nigerian*, 20 October 1979.
104. UNCTAD, *op. cit.*, 45.

capacity, and to ensure a fair distribution of the benefits of economic co-operation. The Declaration provided for the establishment of a full customs union in two stages by 1977 and of a permanent joint commission charged with implementing the agreement. The first phase, to be completed not later than 1 January 1977, included the liberalisation of mutual trade in local products, the harmonisation of rates of import duty and other fiscal incentives, while the second phase would be completed before the end of the same year.

Structurally, the fourteen-point agreement envisages among other things the establishment of a Ministerial Council, with the main responsibility of co-ordinating various committees of experts which are expected to engage in various studies connected with the preparation and establishment of the scheme, and a Secretariat which is the administrative organ of the Union. The latter is directed by a Secretary-General who must always be a Liberian national, assisted by a Sierra Leonean deputy. The headquarters of the Secretariat is at Freetown.

At the centre of the Mano River Union is one key project. The construction of a bridge linking the two countries by road across the Mano River. The African Development Bank approved a loan of some US$1.64 million in 1973 to cover the foreign exchange costs of the project. Its successful completion is expected to increase the present low level of trade between the two countries, as well as facilitiating economic co-operation in other areas. Institutional support has also come from the UNDP. Although steps have been taken to liberalise trade between the two countries, the unofficial export of Sierra Leonean diamonds via Liberia has not stopped. This smuggling costs Sierra Leone dear in both foreign exchange and local government revenue; but the Liberians have much to lose if the illegal trade is stopped. There has been some discussion as to the possibility of the Gambia joining the union, and if this materialises it is easy to foresee a further expansion of the union to the mutual advantage of all.

8(*d*). *The West African Clearing House*. At the meeting of West African central banks held in Accra on 2 and 3 May 1974, a draft clearing agreement was approved. Subsequently, comments on the original draft necessitated a redrafting of the agreement and on 14 March 1975 the definitive text was signed in Lagos by Governors of the Central Banks of the Gambia, Ghana, Liberia, Mali, Nigeria and Sierra Leone and by the Central Bank of the West African states (Benin, Ivory Coast, Niger, Senegal, Togo and Upper Volta). Following the ratification of the agreement the Clearing House came into legal existence on 25 June, 1975 and it started operations on 1 July 1976.

The objectives of the Clearing House as laid down in the Agreement are to promote the use of the members' currencies for intra-regional trade and other transactions, to bring about economies in the use of the members' foreign reserves, to encourage the members to liberalise trade among themselves, and to promote monetary co-operation and consultation.

The basic agreement established three bodies: the Exchange and Clearing Committee, the Executive Secretariat and the Sub-Committee. The Exchange and Clearing Committee, consisting of the governors of the member-banks, carries out extensive functions: it interprets and implements the agreement, determines the transactions to be included or excluded, adopts the rules of procedure, makes regulations for the operation of the Clearing House, and determines the interest rates to be charged on deferred payments as well as the par value of the West African Unit of Account. The Committee meets at least once a year and appoints its own chairman. The Executive Secretary, who is appointed by the Committee, is responsible for co-ordinating, supervising, and controlling the functions provided by the Clearing House. The Sub-Committee, which assists the Committee, meets twice each year on the Committee's advice.

The Clearing House was established in conformity with the provisions of Article 38 of the ECOWAS Treaty. All transactions carried out through it will be expressed in West African Units of Account, into which each contracting party will guarantee the conversion of its currency. At present, the coverage of transactions handled by the Clearing House is too small to exert an appreciable impact on the existing volume and pattern of West African trade. However, it is hoped that its establishment will pave the way for closer monetary integration in the region.

To sum up this chapter, we recapitulate some of the more important factors affecting economic integration in the West African sub-region. Internal political and economic differences, overblown integration schemes (especially so at the initial stages), inadequate physical infrastructure, political instability, external pressures and influences, and pre-occupation with the territorial sovereignty of the nation-state, have all hampered the orderly development of economic co-operation in West Africa. Although the general bases for effective integration in the area are still not very strong, it can surely be expected that, with the increased monetisation and industrialisation of its economies, the opportunities for effective co-operation would progressively increase.

Fortunately, the all-embracing ECOWAS has at last come into existence but the obviously limited success of integration initiatives in the region raises the question whether ECOWAS can overcome the obstacles that have historically plagued earlier integration movements. In the next chapter, the problems and prospects of ECOWAS are exclusively examined against this background.

VI
THE ECONOMIC COMMUNITY OF WEST AFRICAN STATES (ECOWAS): PROSPECTS AND PERSPECTIVES

1. *Origin*

The birth of ECOWAS on 28 May 1975 in Lagos marked the beginning of a new era in the history of economic co-operation in West Africa. Its birth came after a decade of action and inaction. The United Nations Economic Commission for Africa, in Resolutions 142 (VIII) and 145 (VII) passed at its seventh session held in Nairobi in February 1965, recommended that member-states of the Commission should establish as soon as possible sub-regional intergovernmental machinery for harmonising their economic and social development.

Because of the practical difficulties involved in forming economic groupings among independent states, especially less developed ones, it took West African countries ten years from the time of the ECA resolution to form ECOWAS. The period 1965–72 was characterised by hesitation, vacillation and politicking, despite all the institutional support which the ECA gave by organising research and conferences on economic integration in West Africa. In 1972 a series of meetings was initiated by President Eyadema of Togo and General Gowon of Nigeria which, after difficult negotiations, finally brought the community into being. It was these two heads of state who revived the languishing idea and spearheaded the ultimately successful campaign. Both leaders had agreed to form the 'nucleus' of a West African Economic Community which would embrace both Anglophone and Francophone countries. It was feared that the CEAO members would not join the wider regional grouping because of the 'great Francophone-Anglophone divide' in West Africa but surprisingly all embraced ECOWAS. Indeed, eleven heads of state including such influential Francophone leaders as Félix Houphouet-Boigny of Ivory Coast, Ould Daddah of Mauritania and Eyadema of Togo were present when the Treaty was signed in Lagos, confounding earlier speculations that there would be a poor showing, and raising hopes for ECOWAS among the Francophones.

2. *Aims*

The ambitious Clause 65 of the ECOWAS Treaty is expected to

standardise tariffs and trade procedures among the member-countries. Specifically, the sixteen-member community aspires:

... to promote co-operation and development in all fields of economic activity particularly in all fields of industry, transport, telecommunications, energy, agriculture, natural sciences, commerce, monetary and financial questions and in social and cultural matters for the purpose of raising the standard of living of its peoples, of increasing and maintaining economic stability, of fostering closer relations among its members and of contributing to progress and development of the African continent (Article 2[1]).[1]

To this end the Community will *by stages* (emphasis mine) ensure among other things: the elimination of customs duties and other charges of equivalent effect between the member-states in respect of the importation and exportation of goods; the abolition of quantitative and administrative restrictions on trade among the member-states; the establishment of a common customs tariff and a common commercial policy towards third countries; the abolition of the obstacles to the free movement of persons, services and capital between the member-states; the harmonisation of the agricultural policies of the member-states and the promotion of common projects notably in marketing, research and agro-industrial enterprises; the implementation of schemes for the joint development of transport, communication, energy and other infrastructural facilities as well as the evolution of a common policy in these fields; the harmonisation of the economic and industrial policies of the member-states and the elimination of disparities among them in the level of development; the harmonisation, required for the proper functioning of the Community, of member-states' monetary policies; the establishment of a Fund for Co-operation, Compensation and Development; and such other activities calculated to further the Community's aims as the member-states may from time to time undertake in common (Art. 2[2]).

We now examine the organisational structure, the possible benefits, the problems and realities, and the future prospects of ECOWAS in the light of its stated objectives.

3. *Organisational structure*

Article 4 of the Treaty establishes five institutions to run the ECOWAS: the Authority of Heads of State and Government; the Council of Ministers; the Executive Secretariat; the Tribunal of the Community; Technical and Specialised Commissions. The

---

1. See the Treaty of the Economic Community of West African States, Lagos, May 1975, Article 2(1), 7.

Authority of Heads of State and Government is the principal governing institution, responsible for, and with the general direction and control of, the performance of the Community's executive functions; its decisions are binding on all the Community's institutions. The Authority is to meet at least once a year and will determine its own procedures and conduct of business.

Below the Authority of the Heads of State and Government is the Council of Ministers, which will consist of two representatives of each member-state. The responsibilities of the Council of Ministers are to keep under review the functioning and the development of the community in accordance with the Treaty; to make recommendations to the Authority on policy matters aimed at the efficient and harmonious functioning and development of the community; to give direction to all the Community's subordinate institutions; and to exercise other powers conferred on it and perform other duties assigned to it by the Treaty. The decisions and directions of the Council of Ministers, which meets twice a year with a provision for extraordinary meetings, are binding on all the subordinate institutions (Article 6).

The Executive Secretariat, as its name implies, is the executive organ of the Community, and is headed by an Executive Secretary appointed by the Authority to serve for a term of four years, with the possibility of being reappointable for one further term of similar length. According to Article 8 of the Treaty, the Executive Secretary will be the Community's chief executive officer, and he will be assisted by two Deputies appointed by the Council of Ministers. There is also a provision for the appointment of a financial controller and other officers in the Secretariat. The Executive Secretary is responsible for the day-to-day administration of the community and all its institutions.

In view of the complex problems which must arise in the process of harmonising the economies of sixteen different countries with different systems and at different stages of development, the Treaty envisages the establishment of a Tribunal (Article II) to ensure the observance of law and justice in the interpretation of the Treaty's provisions and to settle disputes referred to it.

It also proposes the establishment of four technical and specialised Commissions, namely the Trade, Customs, Immigration, Monetary and Payments Commission; the Industry, Agriculture and Natural Resources Commission; the Transport, Telecommunications and Energy Commission; and the Social and Cultural Affairs Commission — each to have one representative from each member-state. The function of the Commissions is to submit from time to time, through the Executive Secretary to the Council of

Ministers, reports and recommendations in their own respective field of investigation either on their own initiative or at the request of the Council of Ministers or the Executive Secretary. The Treaty can impose additional functions on each Commission.

Other important Treaty provisions include the appointment of an External Auditor of the Community's accounts; the setting up of a Committee of West African Central Banks to oversee the payments system within the grouping; and the establishment of a Fund for co-operation, compensation and development as a mechanism for the equitable distribution of the benefits and costs of integration. The Fund is to be financed from members' contributions, income from community enterprises, external receipts, and subsidies and contributions from all other sources (Articles 10, 38 and 50).

Because of the all-important nature of the Fund as the agency charged with the responsibility for ensuring the equitable distribution of the fruits of the integration experiment, a few comments on its operation may be useful here. Although Article 50 of the Treaty provides for the establishment of the Fund, it is Article 2 of the Fourth Protocol annexed to the Treaty, relating to the Fund, that actually states its purposes:
— to provide compensation and other forms of assistance to member-states which have suffered losses due to the application of the Treaty's provisions;
— to provide compensation to member-states which have suffered losses as a result of the location of Community enterprises;
— to provide grants for financing national or Community research and development activities;
— to grant loans for feasibility studies and development projects in member-states;
— to guarantee foreign investments made in member-states for enterprises established in pursuance of the Treaty's provisions on the harmonisation of industrial policies;
— to provide means to facilitate the sustained mobilisation of internal and external financial resources for the member-states and the Community; and
— to promote development projects in the less developed member-states of the Community.

These are the challenging objectives of the Fund, and their achievement requires not only the active support of the member-states but also that of prospective international investors. In addition to the objectives, the 'Methods of Operation' and 'Operating Principles' are clearly articulated in Articles 10 and 13 of the Protocol relating to the Fund.

Organisationally, the activities of the Fund are regulated and directed by a Board of Directors under Articles 24 and 25 of the Fourth Protocol. The Board is made up of two representatives of each member-state, one of them a member of the Community's Council of Ministers while his alternate director is expected to be a 'person possessing high competence and wide experience in economic, financial and banking affairs'. Article 25(1) vests all the powers of the Fund in the board, which thus takes responsibility for approval of the annual and operational budgets, the making of decisions of a general policy nature subject to confirmation by the Council of Ministers, and approving the administrative policy proposals submitted by the Managing Director. The Board meets at least twice a year and elects its chairman annually from within its own membership. Although it lays down matters of general policy, it does not interfere in the day-to-day running of the Fund's affairs, which is entrusted to the Managing Director, backed by a seventy-two-man team of professionals and other categories of staff, including a Deputy Managing Director. Thus the Managing Director is the Chief Executive Officer of the Fund, although Article 8 of the ECOWAS Treaty requires him to consult the Executive Secretary on certain matters, notably those of crucial significance. Thus there can be no doubt that the Fund is legally and organisationally well-established to fulfil the functions assigned to it by its founding fathers.

However, it has to be admitted that, up till the time of writing, the Fund has not yet financed any project since its inception in 1977, due largely to initial administrative and political teething problems. But the first project to be financed by the Fund was approved by the Authority of Heads of State and Government at their annual summit in May 1980. This is a US$35.5 million telecommunications project, involving eleven international links, seven national links, nine local line plants and thirteen switching centres affecting thirteen of the sixteen ECOWAS states. Fortunately, international interest in the financing of the first ECOWAS project was encouraging: a total financial package of US$69.7 million, almost double the project's total cost, was pledged, and the Fund itself is contributing US$5 million towards financing the project.[2] Despite some initial administrative bottlenecks, the project was expected to take off by the first quarter of 1983.

The ECOWAS Treaty is an ambitious document but also a detailed one. Despite the temptation to rush things, the Treaty

---

2. The authorised capital of the Fund is US$500 million, of which $50 million is the paid-up capital. As at 1 May 1981, $37.5 million of this had been subscribed by the member-states. The Fund can thus borrow or guarantee funds up to the value of its callable capital, namely $450 million.

envisages the gradual achievement of a customs union of West African states over a period of fifteen years.[3] Article 62 specifies that the Treaty and the protocols which may be annexed to it will respectively enter into force provisionally on the signature by Heads of State and Government and definitely on ratification by seven signatory states. The Treaty does not, however, commit the signatories to political or even monetary union. It further binds them not to take measures which may in its own view conflict with the interests of a signatory state. Particularly interesting is the commitment to exchange industrial feasibility studies and information about their industrial experience in order to help minimise the duplication of such studies in the region.

4. Benefits and impact of ECOWAS

A priori, it is generally agreed that trade liberalisation within a grouping maximises economic efficiency from the group's point of view *vis-à-vis* non-trade (i.e. autarky). In ECOWAS, this could be achieved by a variety of ways.

First, assuming that the conditions for effective integration exist, ECOWAS will create a single market of about 145 million consumers in West Africa (see Table II.1, cols 2, 5 and 7), and such a vast market will create opportunities for specialization in patterns of production and for the establishment of large-scale industries through the 'pooling' of national markets. The small size of national markets will no longer act as a brake on the economic development of the 'micro'-states of West Africa, namely Cape Verde, the Gambia, Guinea-Bissau, Liberia, Mauritania and Togo. Indeed, it is estimated that of all the sixteen member-states of ECOWAS only Nigeria and, to a less extent, Ghana and Ivory Coast can at present set up any heavy industry based on the home market;[4] of course, 'heavy reliance on foreign markets is not a sound base for many industries.'[5] It is possible, with resource-oriented specialisation and

---

3. Within a period of two years from the definitive entry into force of the Treaty, member states are to 'freeze' their customs duties and other charges on trade. During the next eight years members will be required to progressively reduce and ultimately eliminate such duties whilst the existing differences in their external customs tariffs would be abolished according to a recommended schedule over the last five years (see Articles 13 and 14). The fourth anniversary of the formation of ECOWAS on 28 May 1979 became the effective date for the freezing of all customs tariffs in all the sixteen-member countries for products originating within the region.

4. See Uka Ezenwe, 'Merits of Integration', *Africa*, January 1975, 23–4; and Uka Ezenwe, 'The Rationale of Economic Integration in West Africa', *Intereconomics*, April 1975, 106–8.

5. S. Kuznets, 'Economic Growth of Small Nations' in *Economic Consequences of the Size of Nations* (proceedings of a conference held by the International Economic Association), London, 1960.

a high level of skilled craftsmanship, for a small country to set up industries that will rely mainly on the export market. The Swiss watch, pharmaceutical and precision engineering industries and the Belgian steel industry are notable examples. But the small countries of West Africa possess neither the resources nor the critical skills required to set up such export-dependent industries. Thus, since rapid industrialisation is the key policy goal of all the members of ECOWAS, the establishment of the grouping must be seen as a curtain-raiser in the drive towards industrialisation.

Secondly, the existence of a wide community market will stimulate production. New industries will be established whilst the existing ones will tend to expand to take advantage of an enlarged market and to exploit the consequent economies of scale. The less efficient intra-union producers, who before integration enjoyed high national tariff protection, will admittedly be hurt by the post-union dismantling of trade barriers; or, in other words, the integration-induced increase in production and the enlargement of market size will generate increased competition, which will probably result in the less competitive industries and/or countries losing markets to more competitive ones within the community, with the consequent benefit of lower prices and a more efficient utilisation of resources. The disruptive effects of such a development on the economies of the members concerned will probably be sharply felt. However, union members are usually permitted under a given union agreement to take the necessary safeguard measures to remedy such a situation. Indeed, Article 26(1) of the ECOWAS Treaty specifically states that 'in the event of serious disturbances occurring in the economy of a member-state following the application of the provisions of this chapter, the member-state concerned shall . . . take the necessary safeguard measures.'

Thirdly, unemployment is one of the most acute economic problems facing West African states today due to the relatively slow rate of industrial expansion *vis-à-vis* the rate of labour supply. Reliable unemployment figures are hard to come by in West Africa, not only because some of the unemployed do not bother to register but also because it is difficult to define unemployment in a predominantly seasonal agrarian setting. However, casual observation shows that the unemployment level is generally high in most countries of the sub-region, and all indications point to the conclusion that the figure is rising. Indeed, as Table VI.1 shows, the average annual rate of labour supply will continue to increase up to the end of the present century. This means increased unemployment and urbanisation with their implications for the provision of social services by individual West African countries. The creation of

ECOWAS has made it possible that, after the transitional period of fifteen years when the customs union would become fully operational, job opportunities may considerably increase following the integration-induced investment in industry. And the proposed abolition of restrictions on the free movement of persons will ensure free mobility of labour within ECOWAS. But while it is true that free mobility of labour within a grouping ensures more efficient utilisation of resources, such a measure often provokes the hostility of the citizens of the host-countries. It should be remembered that in 1965 President Houphouet-Boigny of Ivory Coast made an abortive attempt to introduce dual citizenship within the Council of the Entente.[7] The dual nationality proposal, which was intended to give non-Ivorian migrant workers in the Ivory Coast that country's citizenship while retaining their own individual nationalities, was bitterly opposed by Ivorians, especially the white-collar workers, on the grounds that it constituted a threat to their employment. Thus such a scheme must contain built-in checks and balances if it is to work.

Table VI.1
LABOUR SUPPLY AND URBANISATION IN WEST AFRICA

	Average annual growth of labour force		Urban population as % of total	
	1970-77	1977-2000	1960	1975
Benin	2.1	2.5	10	23
Ghana	2.5	2.9	23	32
Guinea	2.1	2.1	10	16
Ivory Coast	3.9	2.6	19	33
Liberia	2.2	2.4	21	30
Mali	1.9	2.4	11	17
Mauritania	1.8	2.6	3	23
Niger	2.5	2.8	6	10
Nigeria	2.0	2.7	13	18
Senegal	1.7	2.1	23	24
Sierra Leone	1.7	2.2	13	21
Togo	1.9	2.5	10	15
Upper Volta	1.3	2.2	5	8

Source: World Bank, *World Development Report*, 1979.

Fourthly, ECOWAS will promote specialisation and economic efficiency within the union which will in turn bring about improvements in the terms of trade of the integrated group with the rest of the world. Although the terms could be adversely affected, where

---

7. See *Convention on Dual Nationality and Economic Harmonization*, Abidjan, 1965.

increased efficiency in production leads to a higher resource-content per unit of export supply *vis-à-vis* a unit of import, there is no *prima facie* reason why this would be true in ECOWAS. It must be said, however, that the idea of development through trade implies a slow rate of development for the area, given the present small volume of intra-ECOWAS trade. But although intra-area trade accounts for only 3.6 per cent of the member-countries' total trade, it is nevertheless important to the landlocked countries. For instance, the share of their intra-ECOWAS exports to total exports accounts for 55 per cent for Upper Volta, 24 per cent for Mali, and 22 per cent for Niger (Table II.1, cols 10 and 13).

Fifthly, certain industries established in the region operated at less than full capacity due to narrow national markets and/or shortage of inputs. Ghana's Akosombo dam project and Tema refinery, and breweries, textile and cement factories in Ivory Coast are cases in point. Also the planned refineries at Nouadhibou (Mauritania) and Cayar (Senegal), the Kandidji dam (Niger) and the Senengue and Manantali dams (Mali) would need more than their respective domestic markets to operate efficiently. Market integration in the region would clearly lead to the rationalisation and mobilisation of the existing level of excess capacity through vertical specialisation of production processes between plants in the same industry.

Sixthly, ECOWAS envisages the harmonisation of agricultural, industrial and economic policies among its members (Article 2[2]). It is hoped that this will include the standardisation of prices paid officially for agricultural exports. Disparities in agricultural export-import prices have been the major cause of smuggling. In August 1970, while the price of cocoa in Ghana was $292.56 per long ton, the corresponding prices of cocoa in Ivory Coast and Togo were $294.48 and $324.00 respectively. It is therefore not surprising that in the 1970/1 cocoa season Ghana lost 57,000 tons of cocoa through smuggling to her immediate neighbours.[8] Smuggling is also big business elsewhere in the region. It is no secret that Togo exports a quantity of gold which it does not produce; and that transistor radios are smuggled to the Gambia from Senegal while a sizeable proportion of Gambia's imports also illegally find their way into Senegal; and that there is unofficial export of Sierra Leonean diamonds through Liberia. This illegal trade, which has often led to bouts of strained relations between neighbouring West African

---

8. For the costs and effects of smuggling to the economy of Ghana, see A. Kumar, 'Smuggling in Ghana: Its magnitude and Economic Effects', *Nigerian Journal of Economic and Social Studies*, vol. 15 (2), July 1973, 285–303.

states, can be stamped out through joint action by members of the Community.

Seventhly, the joint development of an integrated transport and communications network is another fundamental objective of the newly-formed organisation. The existing network, which was designed during the colonial era to advance the cause of transatlantic commerce, is not adequate for intra-West African trade. The restructuring of the existing transport and communications infrastructure of course can only be done within a regional framework. For example, such interstate projects as the east-west 4,800-km. road linking Dakar and N'Djamena via Senegal, Mali, Upper Volta, Niger, Nigeria, Cameroun and Chad; the West African coastal highway which will link Nouakchott and Lagos via Dakar and Abidjan; and the inter-state road to link Dakar and Bissau via Banjul require the agreement and active participation of all the countries affected. Thus ECOWAS offers the best opportunity for the improvement of the transport infrastructure in the region; and without an adequate regional transport network and communication system, economic integration will have little meaning in a region spread over an area of more than 6,875,000 square km. (Table II.1, col. 7).

Eighthly, the effective operation of ECOWAS can further economic equality not only among member-countries but also among different sectors within a member-country. One of the most important class conflicts in the poor countries of the world today is due to the schism between the poor rural classes and the articulate and powerful urban classes. The view that a common market like ECOWAS can bring about a more equitable distribution of income is based on elements of certain economic theories.[9] One is that rising levels of national income (in this case induced by ECOWAS) will naturally lead to more equality, given the existing free mobility of factors of production. Another argument, based on the classical theorem of factor-price equalisation under free trade, is that increased trade, by reducing disparities between factor prices, will help to equalise incomes. For example, the post-Acheampong deterioration of the economic situation in Ghana forced thousands of skilled and unskilled Ghanaians to take advantage of the ECOWAS free mobility of labour provision (Article 27) to migrate to other countries in the region, e.g. Nigeria, Ivory Coast, Liberia and Sierra

---

9. See Matthew Edel, 'Regional Integration and Income Redistribution: Complements and Substitutes?' in R. Hilton (ed.), *The Movement Toward Latin America Unity*, New York: Praeger 1969, 185–98.

Leone, in search of better employment opportunities (see pp. 190 ff). A third reason, perhaps, is that an integration scheme will create new opportunities which, being new and open to utilisation by new investors and entrepreneurs, will lead to a wider spread of wealth. Each of these effects might develop as here postulated, given the appropriate policy mix, but the reverse effects are also quite possible in the absence of agreed redistributive mechanisms.

Ninthly, the West African drought of 1972–4, affected an area of about 2.5 million square km. This zone generally described as the Sudano-Sahelian zone, includes large parts of Mauritania, Senegal, Mali, Upper Volta and Niger and the northernmost parts of Nigeria. Although the effects of the drought on agricultural production, especially animal husbandry, were more pronounced in this zone,[10] its overall impact was much more widespread. Hence problems posed by such region-wide natural disasters require concerted attack within a regional framework, and it is for such co-operative effort that ECOWAS offers an excellent opportunity. Co-operation among ECOWAS member-states in applied research, improved weather recording and forecasting, the development of river basins, controlled land use and strategic stockpiling would certainly be a more effective means of containing the menace of future droughts than reliance on the *ad hoc* solutions of individual West African countries.

Finally, although ECOWAS — for understandable reasons — did not commit itself to any form of political union, it is anticipated that effective economic integration will in the course of time generate closer political and socio-cultural ties which will in turn reinforce the integration scheme itself. Movement towards economic integration has clearly required an increasing number of decisions to be taken by institutions whose authority must transcend national boundaries to be effective, and this will continue. Consequently, commitment to full economic union implies some form of eventual political union.

However, as experience has shown, West African countries do not want to jeopardise their political independence in the immediate future. Given this attitude, the functional approach to political union should be preferred to the Nkrumah-ist idea of immediate political union as a prelude to economic integration and develop-

---

10. For details see Uka Ezenwe, 'Towards a Regional Policy on post-drought livestock and Meat Production' in *The Aftermath of the 1972–74 Drought in Nigeria*, Federal Department of Water Resources and Centre for Social and Economic Research, Ahmadu Bello University, 1977.

ment.[11] Functionalism originated in 1930s as a system of international peace preservation,[12] and was later elaborated into a general theory of international political integration.[13] For present purposes, a liberal exposition of the theory, based substantially on the contributions of Haas,[14] would be that transnational functional activities induce a gradual, continuous, logical process of politicisation. Politicisation here refers to the supranationalising of the system of legislation, but it also rests upon — and indeed can be defined as — a prevailing loyalty to the central authority. Functional activities are conceived of as technical and therefore non-controversial — generally activities concerned with economic matters: hence the derivation of the theoretical proposition governing the advancement to political union by way of successive 'economic' steps. As to the form of the ultimate political union, a confederal or federal-type structure with a fairly strong executive accountable to a democratically elected parliament is probably the most that can be hoped for. The former, no doubt, could be more easily achieved than the latter. Even so, the attainment of the goal of political union in West Africa through the functional approach has its own limitations. For one thing, the functionalist theory of political integration is a derivation of the experiences of the politico-socio-economic systems of the developed industrial countries; and it embraces a general vision of the functioning of these societies. Thus close examination shows that the existing functionalist idea, not having been modelled specifically for application to West Africa, may not work as anticipated. Indeed, a theory of economic integration, to be adequate, must be specific to the region concerned, and must furthermore be integrated into a general understanding of the functioning of the international system as well as of the specific characteristics of a particular region's integration into the international economy.

5. *Problem areas*

In the foregoing, we have outlined the potential advantages of

11. See Kwame Nkrumah, *Africa Must Unite*, London: Mercury Books, 1965, 150-72, for a brilliant exposition of the idea of African political union. Although the proposal seems dangerously unrealistic, it is nevertheless worth reading.
12. David Mitrany, *A Working Peace System*, London, 1943.
13. E.B. Haas, *The Uniting of Europe*, Stanford University Press, 1958; E.B. Haas, *Beyond the Nation State*, Stanford University Press, 1965; K. Deutsch *et al.*, *Political Community and the North Atlantic Area*, Princeton University Press, 1957; J.P. Sewell, *Functionalism and World Politics*, Princeton University Press, 1966.
14. See E.B. Haas *et al.*, 'Economic and Differential Patterns of Political Integration: Projections about Unity in Latin America', *Journal of Common Market Studies*, June 1967.

ECOWAS to the member-states. But it has also been noted that much depends on the realities of the West African politico-economic setting. We discuss here the more important problems and how they will affect the future development of ECOWAS.

The first problem is that West African countries are predominantly primary producers with the agricultural sector contributing between 23 and 49 per cent of the GDP and employing 50–91 per cent of the labour force (Table VI.2); and their products are oriented to developed markets of Europe and North America rather than those of West Africa.

Ecologically, West Africa has five major crop belts each running from east to west. According to Church[15] the crop belts consist (from south to north) of: (i) rice and tree crops; (ii) cassava, yam, maize and tree crops; (iii) guinea corn, root crops; (iv) millet, groundnuts and cattle; and (v) occasional cereal cultivation by irrigation. This classification has important implications for the volume and direction of intra-West African trade, especially in agricultural products.

First, countries within the same crop belt tend to produce the same or similar agricultural products hence they cannot be important customers to each other. Secondly, most of the industrial goods entering into West African trade are processed agricultural commodities such as sugar, canned beef, frozen meats, tobacco, textiles, leather products, tomato puree, wheat flour and the like and they form a tiny fraction of the total trade. The contribution of industry to GDP as shown in Table VI.2 exaggerates the share of this sector — especially in the case of Nigeria, Liberia, Mauritania and Guinea — because of the importance of mining which is included in the definition of industry in the table. When the definition is restricted to manufacturing activities alone, the relative share of the sector will be much smaller. Thirdly, the flow of West African trade essentially follows the north-south or south-north direction as dictated by the ecological zones, while trade flows from west to east and *vice versa* are relatively small. On the basis of Table II.1 (col. 13), it can safely be concluded that at present Ivory Coast, Mali, Ghana, Upper Volta, Mauritania, Niger, Benin, Nigeria and Senegal are the countries with meaningful trade relations in West Africa.

It is expected that these countries would constitute important 'trade poles' around which intra-area trade will grow in the future. Although the volume of intra-West African trade is hardly more than 3.6 per cent (Table II.1, col. 13) — a factor which minimises

---

15. R.J.H. Church, *West Africa: A Study of the Environment and Man's Use of it*, London: Longmans, 3rd edn, 1961.

Table VI.2
LABOUR FORCE IN AGRICULTURE AND THE DISTRIBUTION OF GDP (%)

	% of labour force in agriculture		Distribution of the GDP					
	1960	1975	Agriculture		Industry		Service	
			1960	1976	1960	1976	1960	1976
Benin	55	50	n.a.	39	n.a.	20	n.a.	41
Cape Verde	n.a.	n.a.	n.a.	n.a.	n.a.	n.a.	n.a.	n.a.
Gambia	n.a.	n.a.	n.a.	n.a.	n.a.	n.a.	n.a.	n.a.
Ghana	64	58	41	49	19	25	40	26
Guinea-Bissau	n.a.	n.a.	n.a.	n.a.	n.a.	n.a.	n.a.	n.a.
Guinea	88	85	n.a.	43	n.a.	33	n.a.	24
Ivory Coast	89	85	43	25	14	20	43	55
Liberia	81	76	40	29	37	37	23	34
Mali	94	91	55	38	10	17	35	45
Mauritania	91	88	57	35	21	37	22	28
Niger	95	93	66	47	10	24	24	29
Nigeria	71	62	63	23	11	50	26	27
Senegal	84	80	30	28	20	24	50	48
Sierra Leone	78	72	n.a.	32	n.a.	23	n.a.	45
Togo	80	73	55	25	16	21	29	54
Upper Volta	92	87	55	34	13	19	32	47

Source: World Bank, *World Development Report*, August 1978.
Note: The Agricultural Sector includes agriculture, forestry, hunting and fishing. The industrial sector comprises mining, manufacturing construction, electricity, water and gas. All other branches of economic activity are regarded as services.

the gains from economic integration — the prospects for the expansion of intra-regional trade are good. It is estimated[16] that by 1990, given the present mix of public policies, the degree of self-sufficiency within the ECOWAS group of countries will be about 47 per cent in wheat; nil in barley; 48 per cent in beef and veal; 55 per cent in mutton and lamb; 49 per cent for other meats; and 81 per cent in fish. The region is expected to be just about self-sufficient in maize, roots and tubers, vegetable oils and poultry.

Production of food crops is surely a potential area for integration of industry and agriculture. It would be not only through food processing and utilisation of the by-products but also through the development of agro-industries based on food crops which would

---

16. See 'A Keynote Address' delivered by Prof. Adebayo Adedeji, Executive Secretary of the UN Economic Commission for Africa, at the Inaugural Conference of West African Economic Association on 'Industrialisation in ECOWAS Countries', Lagos, Nigeria, April 1978.

facilitate the growth of an engineering industry and of food technology in the widest sense including those technologies related to reduction of waste[17] and to packaging and canning.

Another issue turns on transport and communications. As indicated earlier, the present transport and communications systems of West Africa are woefully inadequate and unsuitable for intra-area trade. They display greater heterogeneity and irrationality than, say, in Northern or Eastern Africa. Consider this. Almost 90 per cent of the goods and merchandise of West African countries is transported by sea, and of this traffic 97.5 per cent is carried by non-African shipping lines.[18] As for road transport, not only are the road links sometimes non-existent but where they exist they are frequently not all-weather roads. Apart from the external orientation of African railways, links between them are complicated by differences in specification, while in the case of airlines more than 85 per cent of the subregion's international traffic is carried by non-African airlines; most of the West African airline routes are to Europe and America. Air cargo operations among West African countries are still at a rudimentary stage of development. Many of the airports have no suitable handling equipment (e.g. fork-lifts, high elevators, supply lorries and unloading conveyor belts). In the well-equipped airports, the equipment is often out of service because of poor maintenance. Lack of operational handling equipment causes delays for aircraft: flight schedules are often thrown into confusion at West African airports because of the slowness of the loading and unloading operations. Such delays cause substantial financial losses to the airlines, considering that the cost of a minute's delay of a Boeing is estimated at US$200.[19] An intra-West African telecommunication network is yet to be well developed. Yet poor communications involving time-consuming procedures and insufficient information-sharing can only retard economic growth whereas high intra-ECOWAS transport costs may give 'natural protection' to a number of small-sized plants outweighing the benefits of economies of scale. By definition economic integration between any group of countries implies easy access to each other's market. Therefore, the development of adequate systems for all modern modes of transport should be seen as a *sine qua non* for successful economic integration in West Africa.

17. The issue of reduction of waste is an important matter for West Africa. Consider that a recent ECA study showed that over one-third of the food produced in West Africa is wasted through poor technology, storage and marketing.
18. Ibid.
19. UN, *United Nations Transport and Communications Decade for Africa*, 1978–1988, vol. 2, Part IV (Air Transport), E/CN. 14/726/Add.1, 4.

There is also the question of a common payments system. West African currencies, which are tied to different major hard currencies, are not mutually convertible. Article 38 of the ECOWAS Treaty provides for the creation of a Committee of West African Central Banks which would determine a clearing system of payments for the region. The case for some form of monetary union within the grouping can, of course, hardly be overstated. Currency unification is a very effective answer to the problem of disparate exchange rates within an integration scheme. It would make easier the overcoming of imperfections in commodity and factor movements, and it would allow not only complete freedom of payment by any one place to any other in the union, but it would also give complete freedom to capital movements, which under a common market might still be restricted. It also enhances the opportunities for pooling reserves and makes it easier to pursue a single monetary policy. Thus currency unification takes care of the problems of balance of payments in intra-union trade — enabling surplus countries to avoid inflationary policies and deficit countries to avoid deflationary policies. However, currency union is a very advanced form of integration, and the obstacles to its formation among beginners are formidable.[20] In the first place, certain necessary pre-conditions — e.g. factor mobility, increased intra-regional trade, and harmonisation of financial policies — are non-existent.

It was against this background that the ECOWAS Secretariat commissioned a study to 'assess the possibilities of achieving limited convertibility with specific reference to prospects for liberalisation and harmonisation of exchange controls and restrictions, taking into account the long-term objective of monetary union within the ECOWAS Region'. The Report,[21] which was prepared and submitted by an IMF team in October 1980, considered three exchange rate options under which convertibility of currencies could be achieved. These were, first *a full currency union* involving a common currency and a regional monetary agency; secondly *a semi-full currency union* involving the maintenance of individual currencies but the setting up of irrevocably fixed exchange rates between them, and thirdly *an exchange agreement* whereby the currencies of all member-states will be exchanged at rates to be deter-

---

20. For further reading, see E. Osagie, 'Monetary Disintegration and Integration in West Africa', *Nigerian Journal of International Studies*, July 1975, 19-27, and R.F. Kahn *et al.*, 'The Contribution of Payments Arrangements to Trade Expansion' in P. Robson (ed.), *International Economic Integration*, Harmondsworth: Penguin Books, 1971, 242-53.

21. IMF, *Convertibility in the Economic Community of West African States*, Lagos, October 1980.

mined by the cross-rates between the currencies of all member-states and the rate prevailing between each currency and that of its reference currency. When Governors and Directors of Research of West African Central Banks met in Accra in May 1981 with ECOWAS and IMF officials in attendance, the consensus of opinion among the participants was that under existing circumstances an exchange agreement would be the most realistic avenue for monetary co-operation within ECOWAS.

The exchange agreement as envisaged takes into account the realities of the situation (the varieties of exchange regimes) existing in the sub-region. It allows for freedom of choice as to the type of exchange rate arrangement to be adopted by each country. In addition, since exchange rates would not be irrevocably fixed under this arrangement, they would be free to move — reflecting, among other factors, the absence of harmonisation in the monetary and fiscal policies of member-states of the Community. The other two options, which would necessarily have entailed total or partial loss of autonomy in monetary and fiscal policies, were ruled out as unacceptable to individual members.

Having agreed on the appropriate exchange arrangement for ECOWAS, the next question is how to bring it about. On this the Executive Secretariat prepared a programme for monetary co-operation which envisages the conclusion of the exchange agreement by 1989. The programme,[22] which is linked to the trade liberalisation plan of the Community, is as follows:

— agreement on liberalisation of exchange controls on current transactions within the Community, to be reached by May 1983;
— agreement to be reached for the setting-up of a balance of payments support fund, also by May 1983 (the support fund could be linked with the operations of the West African Clearing House);
— agreement to be reached on the liberalisation of exchange controls on capital transactions within the Community, May 1986;
— agreement to be reached on the harmonisation of monetary and fiscal policies of member-countries, May 1988;
— the coming into existence of an exchange agreement, May 1989.

This impressive and optimistic time-table represents, as the author observed, the genuine wishes of the participants at the meeting of experts that considered the exchange proposal. However, up to the time of going to press, the first two agreements have not been realised and no new dates have been fixed, ostensibly because member-states are still studying the effect of the proposed measures on their

---

22. ECOWAS Secretariat, 'A Programme for Monetary Co-operation' (official working paper), Lagos, 1981.

economies. Thus, although the prospects are good for realising the ECOWAS exchange rate programme, it is doubtful whether progress will be fast or smooth.

Meanwhile, some encouraging progress has been made in other aspects of monetary co-operation in the sub-region. The establishment of the West African Clearing House,[23] which came into legal existence on 25 June 1975 and started operations on 1 July 1976, has paved the way for closer monetary integration in West Africa. With the establishment of the first region-wide Clearing House in West Africa, it is expected that the existing bilateral arrangements in the region (such as those between Ghana and Mali; Upper Volta and Togo; Niger and Benin; Nigeria and Niger; and Togo and Mali) would end as soon as they expire, thereby giving way to smooth flow of trade and payments throughout the whole region.

However, this is no more than a hope. To begin with, membership is open to all West African countries but it is optional. Mauritania and Guinea are yet to join the clearing arrangement, although their membership of ECOWAS will eventually compel them to do so. Furthermore, the coverage of transactions handled by the clearing house was considered within the framework of the peculiarity of trade in the sub-region, namely the limited volume of intra-area trade, the narrow composition of trade items, especially with some countries relying on only one or two commodities, widespread unofficial frontier trade transactions not easily amenable to official control and the movement of goods which do not originate in the region. The net effect of these structural problems is to limit the effectiveness of the West African Clearing House as a mechanism for rapid intra-West African trade.

The determination of the appropriate level of common external tariff is another issue that may not be easy. In fixing the level, consideration must be given to the need to protect the community's industry and to compensate its members for loss of revenue arising from the abolition of duties inside the community. A low tariff rate would negate these considerations while a high one would overprotect intra-union producers, thereby encouraging inefficiency. Since both objectives are quite legitimate, a trade-off between them has to be worked out. Apart from the height of the common external tariff, there is the question of the implementation period. The weaker member-countries may require a longer time than others to absorb the shock stemming from premature tariff disarmament. Articles 13 and 14 of the Treaty stipulate a two-year tariff grace

23. See *The West African Clearing House: an Experiment in Multilateral Co-operation in West Africa*, Freetown, 1976.

period, another eight-year tariff disarmament span, and a further five-year common external tariff establishment span for all members. Although the fifteen-year period of progressive tariff disarmament and the establishment of a common external tariff for the entire community appear reasonable from the group standpoint, the Treaty failed to make any special provision for the weaker economies. Certainly, the poorer members would require more time to make the painful adjustments prompted by a new tariff regime.

Furthermore, the provision to ensure that all states have a similar 'industrial climate' to make for smoother industrial development is admirable. Yet it is difficult to imagine how this can be easily achieved in the short run in a region following different ideological paths to national economic development, especially between the undisguised capitalist mode of Ivory Coast and the thoroughgoing socialist system of Guinea, which has now been joined by Benin. The others emphasise — in varying degrees — the desirability of capturing the commanding heights of their national economies; and indigenisation and nationalisation of the strategic economic sectors have been recognised as the appropriate policy tools for achieving this goal.

The achievement of a common industrial climate in West Africa should be seen as a long-term problem and would necessarily require profound political and ideological changes. However, though desirable, it need not be a pre-condition for economic integration, provided a measure of working harmony can be effected. So long as member-states retain their political sovereignty, the investment climate would normally reflect national priorities and goals, which, for good reasons, may differ from country to country.

The biggest danger facing ECOWAS is probably the existence of other rival integration schemes in West Africa. Aware of this and desirous of accommodating them, the Treaty allows member-states to belong to other economic groupings provided that such membership does not 'derogate from the obligations of that member-state under this Treaty' (Article 20[3]). But there are other problems. Obviously, such other organisations which appear to be 'complementary' by nature, like the Organisation for the Development of the Senegal River (membership: Mali, Mauritania and Senegal) or the Mano River Union between Sierra Leone and Liberia, or even wider bodies like the ECA can only help to develop the new Community. But the incorporation of the Francophone Economic Community of West Africa (CEAO), whose members are Ivory Coast, Mali, Mauritania, Niger, Senegal and Upper Volta, may be less easy. For this six-member body is essentially a miniature ECOWAS of the Francophones. Indeed, it is feared that the French-inspired CEAO sees itself as a rival to ECOWAS, for which reason CEAO invited the Anglophone West African countries to join it if they so wished. None has so far taken up the offer. In November 1978, for instance, while the ECOWAS ministerial meeting was taking place in Dakar, CEAO was holding a simultaneous summit in Bamako. One can only hope that since the long-term aims of the two rival organisations are basically the same, some sort of merger will be worked out. Otherwise the simultaneous operation of the two rivals will make both ineffective.[24]

However, because of the potential danger inherent in this matter it is receiving priority attention. In May 1980, CEAO applied to ECOWAS for derogation with respect to the implementation of the provisions of Article 20 of the ECOWAS Treaty. The problem of derogation stems from the fact that in 1976 CEAO instituted a tariff regime for the benefit of its members, involving a preferential taxation relating to the import of non-processed goods — a system involving total exemption from entry duties and taxes — and the operation of a preferential regime of regional co-operation tax, which contributes to the Community Development Fund. Since CEAO members are also members of ECOWAS, CEAO is required under the ECOWAS Treaty (Article 20) to extend 'most favoured nation' treatment immediately to other ECOWAS member-states. But the CEAO members feel that their intra-CEAO tariff regime is more favourable than that of ECOWAS, and would neither abolish it nor extend it to other ECOWAS members at this early stage. Earlier, the CEAO heads of state had unanimously passed a resolution 'to request the ECOWAS authorities to grant them a seven-year period, starting from the date of the implementation of the ECOWAS programme on the elimination of tariffs, to retain in their trade the preferential system currently in force in CEAO, without automatically extending it to other ECOWAS member-states'.

Realising that it needed similar exemption, the Mano River Union also applied to ECOWAS. Faced with a knotty issue that could cripple the young organisation, a seven-nation ECOWAS ministerial committee was set up in May 1981 to study the joint request and its implications and make recommendations to the Authority of Heads of State of ECOWAS. When the Committee reported six months later, it strongly recommended a comprehensive study the

---

24. Members of the Community appear to have seen the writing on the wall; hence they are working hard to avert the disruptive effects of the existence of these rival organisations (see The Proceedings of the 9th and 10th Sessions of the ECOWAS Council of Ministers, 1981).

objective of which would be the convergence of the three systems — ECOWAS, CEAO and MRU.

The first part of the study reviews the trade liberalisation mechanisms of the three organisations and the compensation mechanisms. Once the first phase of the study is completed, the length of the adjustment period for the organisations (surely necessary to avoid disruptive effects from too sudden change to a new system) would be fixed. The second part of the study would focus on harmonisation of the internal taxation of member-states; establishment of a common external tariff; and the establishment of supporting measures for the trade liberalisation programme: in sum, the definition of a coherent economic development policy for the sub-region. It is proposed to complete the second phase of the study before the end of the adjustment period. During the intervening study period, the Community resolved to take the following measures:

— To allow the simultaneous application of the three systems: CEAO and MRU to apply their internal regulations among their respective member-countries; in their relation with other ECOWAS member-countries, CEAO and MRU countries to apply ECOWAS regulations;

— To work towards the consolidation of the current rates and lists of products that consitute the CEAO and MRU trade preferential systems. With respect to the admission of new products, the two organisations to apply ECOWAS rules; and

— To insist on the application of the Customs and Statistical documents of ECOWAS as from 1 January 1982.[25]

The determination of each group to retain its smaller organisation while belonging to the all-embracing ECOWAS indicates that if these compromise measures can be successfully implemented, they will go a long way towards accommodating interests that seem to conflict with one another.

Politically, there is another dimension to this internal rivalry. France appears frightened of losing its dwindling hold over its ex-colonies where billions of francs have been invested. The success of ECOWAS will surely weaken this hold still further — in the way that Nigeria's apparent success in its initial mediation efforts between the two opposing factions in Chad appeared to do at the time. France is therefore supporting the CEAO through its satellite states as a bulwark against ECOWAS. Thus the influence and

---

25. See Proceedings of 9th and 10th Sessions of ECOWAS Council of Minister, 1981.

economic interests of non-African countries may eventually retard the orderly progress of ECOWAS.

More recently, the often dreaded but not totally unexpected problem of non-implementation of decisions and non-fulfilment of obligations on the part of member-states seems to have assumed disturbing dimensions. Empirical evidence shows that non-compliance is the principal canker within international organisations, and the survival and effectiveness of any such organisation depends largely on the co-operation and sense of commitment of its members.

The issue of non-compliance among ECOWAS members has two aspects: first, monetary contributions do not flow regularly as scheduled and, secondly, the ratification and implementation of decisions reached are not strictly adhered to. As regards financial contributions, the random pattern of payments is illustrated in Table 6.3. On 30 April 1981, only one country had paid part of its contribution to the operational budget of the Executive Secretariat for 1981, and only seven countries had made contributions to the 1980 budget — four in full and three in part. Seven countries' contributions were still outstanding for the 1979 budget; three for the 1978 budget and one for 1977. The last-mentioned country had, up to that date, made no contribution whatever to any one of the Secretariat's budgets. A total of US$12,133,160.98 was still in arrears.

The situation in the Fund is essentially the same. Contributions to its operational budgets for 1977 and 1978 have not been paid in full by all the member-states; and it is clear from Table VI.3 that, as at 1 May 1981, 25 per cent of the Fund's paid-up capital was still in arrears. Despite these outstanding contributions, however, the Fund — unlike the Executive Secretariat — does not face any financial crisis at this point in time. There are two main reasons for this. Firstly, up to the time of writing, ECOWAS has not financed any development project; and secondly, it has been earning enough money from its portfolio investments since 1980 to be able to finance its annual budgets without relying on annual contributions. The position of the Executive Secretariat is quite different. At this early stage in the Community's life the Secretariat depends exclusively on annual budgetary contributions from member-states — a factor which explains why sometimes frantic efforts have to be made to raise enough money for staff salaries and consultancy charges.

On the issue of implementation of decisions and protocols, progress has also been poor. At each meeting of the Authority of Heads of State and Government since April 1978, a number of protocols have been signed and decisions taken but the ratification

Table VI.3
AMOUNTS OWED BY ECOWAS MEMBERS TO THE SECRETARIAT, 30 April 1981
(in Units of Account)

	1977	1978	1979	1980	1981	Total
Benin	—	—	—	—	193,708.95	193,708.95
Cape Verde	—	—	—	68,469.00	64,569.65	133,038.65
Gambia	—	—	103,598.00	178,027.00	167,881.09	449,506.05
Ghana	—	—	—	389,618.34	832,548.45	1,222,566.83
Guinea-Bissau	—	—	69,623.00	102,703.00	96,854.48	269,180.48
Guinea	—	—	34,630.00	198,557.00	187,251.99	420,438.99
Ivory Coast	—	—	—	—	80,177.89	80,177.89
Liberia	—	122,576	312,150.00	458,736.00	432,616.65	1,326,078.65
Mali	58,900	34,673	88,527.00	130,119.00	122,682.33	434,501.23
Mauritania	—	36,621	167,832.00	246,484.00	232,450.74	683,587.74
Niger	—	—	—	—	135,596.26	135,596.26
Nigeria	—	—	—	—	2,117,884.52	2,117,884.52
Senegal	—	—	229,865.00	369,727.00	348,676.11	947,266.11
Sierra Leone	—	—	—	250,345.00	284,106.46	534,451.46
Togo	—	—	—	187,852.00	232,450.74	420,302.74
Upper Volta	—	—	—	149,534.00	167,881.05	317,415.09
Total	58,900	194,070	1,005,225.00	2,724,171.34	5,697,737.37	9,680,103.37

Note: The average conversion rate used is 1UA = US$1.28.
Source: Official data.

Table VI.4
SUBSCRIPTION BY ECOWAS MEMBERS TO THE FUND'S PAID-IN CAPITAL, 30 April 1981

(1)	Ratio (2)	Allocation ($) (3)	Paid-in capital ($) (4)	(4) as % of (3)
Benin	3.0	1,500,001	1,500,001	100.00
Cape Verde	1.0	500,000	250,000	50.00
Gambia	2.6	1,300,000	250,025	19.23
Ghana	12.9	6,449,979	1,034,312	16.03
Guinea-Bissau	1.5	750,000	713,329	95.11
Guinea	2.9	1,450,097	1,450,097	100.00
Ivory Coast	13.0	6,499,978	6,346,346	97.63
Liberia	6.7	3,350,000	1,644,078	49.07
Mali	1.9	950,000	5,052	0.53
Mauritania	3.6	1,800,000	—	0.00
Niger	2.1	1,050,000	1,050,000	100.00
Nigeria	32.8	16,399,945	16,259,937	99.14
Senegal	5.4	2,700,000	2,700,000	100.00
Sierra Leone	4.4	2,200,000	2,200,000	100.00
Togo	3.6	1,800,000	1,800,000	100.00
Upper Volta	2.6	1,300,000	335,472	25.80
Total	100.00	50,000,000	37,538,649	75.07

Note: Outstanding contributions to the Capital stands at US$12,461,350.67 which is due from ten member-states, although Guinea-Bissau, Ivory Coast and Nigeria appear to have transferred in full their respective allocations to the Fund but exchange rate variations affected the actual amount received.
Source: Official figures.

of these protocols has been slow. Table VI.5 demonstrates that, up to end of May 1981, only four member-states had ratified the Protocol on Non-Aggression and the Protocol on Privileges and Immunities, both signed in 1978. Only nine member-countries had ratified the Protocol on Free Movement of Persons, Right of Residence and Establishment; and only five had ratified the supplementary Protocol of Rectification of the French text of the Protocol relating to the Definition of the Concept of Products originating from member-states of ECOWAS, and the Supplementary Protocol amending the Protocol on Originating Products, all three of which were signed in May 1979. No member-state has yet ratified the Supplementary Protocol amending Article 8 of the French text of the Protocol relating to the Definition of the Concept of Originating Products signed in May 1980. In other words, no member-state had ratified all the protocols signed by the Authority since 1978. Six members had ratified *none* at all, two had ratified four, five had

Table VI.5
ECOWAS MEMBERS: DATES OF RATIFICATION OF PROTOCOLS

	Non-Aggression Pact	Privileges and immunities	Free movement of persons	Rules of origin (French text)	Amending rules of origin	Supplementary rules of origin
Benin						
Cape Verde						
Gambia						
Ghana	30.3.79	30.3.79	31.10.80			
Guinea-Bissau		27.4.79	8.4.80			
Guinea			17.10.79	20.8.79	8.1.80	
Ivory Coast			20.5.80	17.10.79	17.10.79	
Liberia			1.4.80	14.2.80	14.2.80	
Mali						
Mauritania						
Niger	17.5.79		11.1.80	11.1.80		
Nigeria	18.5.79		12.9.79	12.9.79	12.9.79	
Senegal		28.2.79	24.5.80			
Sierra Leone						
Togo	3.3.80	3.3.80	9.11.79		3.3.80	
Upper Volta						

Source: Official records.

ratified three, one has ratified two, and two had ratified only one.

Because a protocol comes definitively into operation only when a minimum of seven member-states have ratified it, only one of the six protocols signed by the Authority since 1978 — that on Free Movement of Persons — has so far become fully operational. Even in the case of this particular protocol, Liberia, which has ratified it, now finds herself unable to implement it for reasons of security. Undoubtedly, the non-ratification and non-implementation of these protocols create a serious obstacle to the Community's smooth development.

Finally, ECOWAS represents another glaring case of a partnership of unequal partners. In theory, economic integration is likely to be more successful where the members are at a comparable stage of development or have a small industrial base and/or where none of the members is so large as to permit it to pursue a national industrialisation programme independently as an alternative to integration. Nigeria, one of the sixteen members of ECOWAS, can to a large extent go it alone. Endowed with rich natural resources, including petroleum, it has about 57 per cent of the population of the area, and its GNP accounted for 69.2 per cent of the sub-regional total in 1979 (Table II.1). It has also the largest and most diversified industrial sector, at least in absolute terms. Given the tendency of industries to cluster in a few industrial growth points, there is some fear that ECOWAS might widen rather than narrow the 'economic gap' between its members. This is perhaps the reason why Senegal was reluctant to ratify the Treaty of Lagos. Be that as it may, there is no reason why a satisfactory distributive mechanism cannot be built into the ECOWAS Treaty to avert the danger of domination and unfair distribution of the benefits and costs of integration.

The difficulties and problems outlined above are by no means peculiar to ECOWAS. In varying degrees, they are common to almost every integration movement in the Third World, be it the Central African Customs and Economic Union (UDEAC), the defunct East African Common Market (EACM), the Arab Common Market, the Latin American Free Trade Area (LAFTA), the Central American Common Market (CACM), the Caribbean Free Trade Association, and so on. The known dangers should spur all the parties concerned to greater efforts and serve as a warning against excessive optimism over what the present endeavour can achieve in the short run.

6. Realities and outlook

West African countries, like other parts of the developing world, have, at least conceptually, two alternatives for the fostering of their economic growth. The first is the autarkic path. This, however, can be dismissed quickly, given the high costs it imposes and, more important, its particular unsuitability for West African countries. The second alternative, the so-called 'outward-looking' path of development, in which exports are concentrated in a few traditional products, also does not offer good prospects. The principal reasons for this are well-known. With few exceptions, the export products of less developed countries (LDCs) are very unstable; the economic growth of countries following the foreign-trade-oriented path of development depends heavily on economic conditions in the industrial countries which they cannot influence; and finally the industrial centres — despite the talk of a new international economic order — have established so many obstacles to trade and migration that it is very difficult for LDCs today to obtain the benefits of their comparative advantages.

It is against this sombre background that most LDCs have been seeking alternative strategies for economic development; and economic integration becomes an alternative strategy, for it represents a middle way between the autarkic and the 'outward-looking' paths of development. It takes something from both of them: from autarky the protection of the integration area from outside competition through the common external tariff, and from the outward-looking path of development the opening of the national markets of each of the member-countries to regional competition. Thus the integration programmes involving LDCs should be seen much more as an alternative path of economic development than as a way of allowing a better allocation of a given stock of factors of production or as a mere tariff issue.

Even so, it is necessary with most LDCs to avoid expecting too much from economic integration, which cannot be a substitute for the internal efforts and measures that each country has to take. In Castillo's words, 'economic integration is not a way to evade reforms by exporting the problems to the rest of the region.'[26] Individual West African countries are in urgent need of internal reforms and structural transformation of their economies with a view to ensuring the creation and optimal use of factors of production. Countries of the region must tap and exploit their resources to the full. Non-African multi-national corporations

---

26. C.M. Castillo, *Growth and Integration in Central America*, New York: Praeger, 1965, 113.

operating in West Africa should be replaced by indigenous multinational firms, if a large part of the benefits of integration is to be retained in the region. But this cannot be done by decree, West African businessmen must take up the challenge. Furthermore, internal economic policies pursued by individual members of ECOWAS must be consistent with the integration idea. For example, the indigenisation policies of Ghana and Nigeria failed to take into consideration Article 27 of the ECOWAS Treaty, which provides for 'freedom to reside and to work anywhere within the community'. Similarly, other incompatible monetary, fiscal and budgetary measures should be aligned with those of the community.

Serious efforts should also be made to obliterate the remaining vestiges of the colonial legacy in West Africa in the relations between the Anglophones and Francophones. It seems that what is now regarded as 'political differences' between the Former President Senghor of Senegal, the unofficial mouthpiece of the Francophones, and Nigeria, the spokesman of the English-speaking countries, is traceable to past cleavages. Senegal sees ECOWAS as a possible channel through which Nigeria, given its economic and political weight, could expand its wings over Africa south of the Sahara and assume a role as 'Big Brother';[9] hence, it is thought, Nigeria's particular eagerness to see ECOWAS succeed. French governments, working on this seemingly false premise, have consistently supported and encouraged the strengthening of CEAO for reasons of self-interest. If there is concern in Senegal about domination by Nigeria within ECOWAS, and in France about its own declining influence over its ex-colonies, then there would be a common interest among them in inhibiting the development of ECOWAS. The fear of Nigerian domination is unreal, and so this would serve no useful purpose.

But if the all-embracing West African integration experiment that is ECOWAS is to have a chance of success, six conditions will have to be met:

(i) member-countries should align their internal development efforts and policies with those of the community;
(ii) the community should always enjoy strong political support from member-states, irrespective of which governments are in power;
(iii) there should be commitment to the integration idea and its success among all groups in the region, whether social, professional or political; among private enterprises, and among households and individuals;
(iv) member-governments must accept a supra-national authority whose decisions are binding on all of them;

(v) members should honour all obligations (e.g. making their statutory contributions) promptly and regularly; and
(vi) the community must establish a fair and acceptable distributive mechanism to ensure equitable distribution of the benefits and costs of integration.

As the lessons from other extant or defunct integration schemes have shown, failure to satisfy one or more of these conditions could land ECOWAS in serious trouble.

VII
THE ROLE OF EXTERNAL FORCES

We have tried to show in the preceding chapters the progress and potential of economic integration in West Africa and the problems facing it. In this chapter we want to look briefly at the effects of external forces on co-operation in the region with a view to making some policy recommendations.

1. What external forces?

For a developing area like West Africa, which still depends largely on the importation of capital, technology and skills from the rest of the world for its development, the international environment is a very important variable in the integration development process. *Ipso facto*, this potentially precarious situation places the region at the mercy of powerful external forces and influences. The external forces implied here include foreign governments, national and international aid agencies, private foundations and multinational corporations. By acting either as 'catalytic' agents or in some other role, these external forces can influence the integration process in a variety of ways.

First, agencies can form consortia to pool resources to be placed at the disposal of integration schemes. Depending on how long the assistance lasts, it can contribute to the equitable distribution of the gains from integration by funding compensatory agreements. This is particularly so since the 'richer' members of a grouping generally do not feel responsible for the 'poorer' ones; the poverty of the latter may be a result of historical conditions, the random distribution of natural resources or past policies of extra-regional agents. Unhappily, the poorer members see the situation differently: the misfortunes of the present are added to those of the past, and can easily be associated with the unjust functioning of the grouping, for which someone has to be blamed. Until the collapse of the East African Common Market its ills were often blamed on the relatively most prosperous member, Kenya.

Secondly, foreign aid donors can help a new integration scheme to get off the ground by providing institutional support when it is most needed; and the institution so established may need technical assistance for a considerable length of time until it can stand on its own. For instance, ECOWAS has received a lot of technical support from the ECA and UNCTAD since its inception. Integrative schemes can be effectively encouraged in this way.

Thirdly, given the fact that foreign bilateral aid is not altogether an unmixed blessing, multi-lateral aid may sometimes be more appealing to aid recipients because the element of self-interest may play a less important role for the donor. Even so, such aid donors sometimes prefer to finance common or regional projects of an infrastructural nature. The African Development Bank, like other such banks, accords special priority to projects or programmes whose nature or scope concerns several members, and which are intended to make the economies of its members increasingly complementary. In December 1965 the Bank's Board of Governors agreed that the agency would finance particularly those projects that are included within regional development programmes, and that special preference would be given to projects that might benefit two or more member-countries and thus stimulate inter-African co-operation.

Fourthly, the regionalisation of aid could increase the overall performance of economic aid where the total flow of external aid is not expanding *pari passu* with the needs of the region. And finally, the co-ordination and regionalization of external assistance can minimise the wasteful duplication of certain investments, feasibility studies and research expenditure and prevent the adoption of 'beggar-my-neighbour' policies which run counter to regional integration motives.

However, it is widely acknowledged that foreign aid, rather than being an altruistic means of assisting the poor, has often been used to promote the political and strategic interests of the donor or to secure the greater dependence of the recipient on the latter. This notwithstanding, it is important to look at the specific impact. Of all the external forces or organisations interested in economic integration among LDCs the roles of the United Nations system and the transnational firms are by far the most profound. We shall now briefly consider them.

1(*a*). *The United Nations system.* The UN system plays a unique role in fostering economic integration in different parts of the developing world. This is reflected in the number of projects currently being financed by the UN agencies, especially in West Africa. Indeed, the Seventh Special Session of the UN General Assembly called on the UN system

> to provide, as and when requested, support and assistance to developing countries in strengthening and enlarging their mutual cooperation at subregional, regional and interregional levels. In this regard, suitable institutional arrangements within the United Nations development system should be made and, when appropriate, strengthened, such as those within the United Nations Conference on Trade and Development, the United Nations

Industrial Development Organisation and the United Nations Development Programme.[1]

Other arms of the UN system have also been actively involved in the process of economic integration in West Africa, and have helped in various ways to advance its cause, especially the ECA, WHO, FAO and UNESCO (Table VII.1) — despite their different functions. This chapter examines the nature of the assistance that has been provided in the past.

Of course, the UNDP is the leader of the UN development activities and its resources tend to overshadow those of any other UN organisation. Table VII.1 bears eloquent testimony of this: within a space of four years, 1972-6, it funded well over forty-five important projects. It is clear from the table that the UNDP's activities are consistent with its avowed policy as enunciated in the following major policy statement issued in 1975:

Projects for regional and subregional integration, for technical and institutional support for the harmonization of policies among developing countries, for evolving a consortium approach to assistance involving mobilization of the resources of bilateral and other multilateral programmes and for co-ordinated use of the major resources of developing countries within the framework of multilateral co-operation such as the Mekong project, the integrated development of the Senegal, Niger and Kagera river basins etc. are other examples of activities on which UNDP might well concentrate by virtue of its unique advantages and in view of the larger objective of the United Nations system to promote international good will and harmony.[2]

It is estimated that in 1976 the UN system committed well over US$67 million to the financing of projects in West Africa. In its regional programmes its priorities have always been in the following order: (i) economic integration; (ii) co-operation on the joint solution of common problems such as river basin development; (iii) joint education ventures in branches of study where either the number of students was small, as in telecommunications, postal work, civil aviation and meteorology, or overheads were very high; (iv) inter-country activities aimed at overcoming the shortcomings in inter-country communications; (v) projects designed to solve research problems which have arisen in more than one country and for the solution of which joint efforts should be organised; (vi) operations fostering intra-African trade, and trade between African countries

---

1. See General Assembly Resolution 3362 (S-VII), section VI, para. 1.
2. J.P. Renninger, *Multinational Co-operation for Development in West Africa*, Oxford: Pergamon Press, 1979, 75.

Table VII.1
SOME RELEVANT UNDP INTER-COUNTRY PROJECTS IN WEST AFRICA, 1972–76

Project	Countries covered	Executing agency
Flood control and warning system in the River Niger basin (Phase 2)	Guinea, Mali, Niger	WMO
Hydrological forecasting system for the Middle and Lower basins of the Niger River	Benin, Cameroon, Mali, Niger, Nigeria, Upper Volta	WMO
Programme for the strengthening of agrometeorological and hydrological services of the Sahelian countries: training centre for agrometeorology and applied ayrology	Chad, The Gambia, Mali, Mauritania, Niger, Senegal, Upper Volta	WMO
Onchocerciasis control in the Volta River basin	Benin, Ghana, Ivory Coast, Mali, Niger, Togo, Upper Volta	WHO
Onchocerciasis control programme in the Volta River basin area: Applied research (epidemiology and chemotherapy) and training	Benin, Ghana, Ivory Coast, Mali, Niger, Togo, Upper Volta	WHO
Applied research on trypanosomiasis epidemiology and control	Ivory Coast, Niger, Upper Volta	WHO
Regional centre for postal training, Abidjan	Ivory Coast, Mali, Mauritania, Niger, Senegal, Upper Volta, Togo	UPU
Documentation centre for the Niger basin	Cameroun, Chad, Benin, Guinea, Ivory Coast, Mali, Niger, Nigeria, Upper Volta	UNESCO
Telecommunications network adviser (W. Africa)	West African Sub-region	ITU
Multinational school for medium–level telecommunications personnel, Rufisque	Benin, Ivory Coast, Mali Mauritania, Niger, Senegal, Upper Volta	ITU
Telecommunications link between the Gambia and Senegal	The Gambia, Senegal	ITU
Hydroagricultural survey of the Senegal River basin (OMVS)	Mali, Mauritania, Senegal	FAO
Water resources, Lake Chad basin	Cameroun, Chad, Niger, Nigeria	FAO
Water resources in the Lake Chad basin (LCBC)	Cameroun, Chad, Niger, Nigeria	FAO

What external forces?

Project	Countries covered	Executing agency
Agricultural research and its application in the Senegal River basin (OMVS)	Mali, Mauritania, Senegal	FAO
Documentation centre for OMVS	Mali, Mauritania, Senegal	FAO
Livestock development in Assale-Servewel (LCBC)	Cameroun, Chad, Niger, Nigeria	FAO
Four agricultural centres in the Lake Chad basin	Cameroun, Chad, Niger, Nigeria	FAO
Development of fisheries in Lake Chad	Cameroun, Chad, Niger, Nigeria	FAO
Research on desert locust (OCLALAV)	Benin, Cameroun, Chad, Ivory Coast, Mali, Mauritania, Niger, Nigeria, Senegal, Somalia, Upper Volta	FAO
Control of grain-eating birds (Phase 2)	Benin, Cameroun, Chad, Ivory Coast, Mali, Mauritania, Niger, Nigeria, Senegal, Somalia, Upper Volta	FAO
Agricultural development in the Senegal River basin (Phase 2) (OMVS)	Mali, Mauritania, Senegal	FAO
Hydraulic development of pastoral areas (LCBC)	Cameroun, Chad, Niger, Nigeria	FAO
Creation of four forestry centres around Lake Chad (LCBC)	Cameroun, Chad, Niger, Nigeria	FAO
Implementation of water drilling programme in the Lake Chad basin (LCBC)	Cameroun, Chad, Niger, Nigeria	FAO
Bornu livestock extension (LCBC)	Cameroun, Chad, Niger, Nigeria	FAO
Applied research on tsetse control in dry savanna zones	Ivory Coast, Niger, Nigeria, Upper Volta	FAO
Study of the Mano River basin: land resources survey (Liberian portion)	Liberia, Sierra Leone	FAO
West Africa rice development association (Phase 1 and 2)	Benin, The Gambia, Ghana, Ivory Coast, Liberia, Mali, Mauritania, Niger, Nigeria, Senegal, Sierra Leone, Togo, Upper Volta	FAO
West African Clearing House agreement	Benin, Cameroun, Ghana, Ivory Coast, Liberia, Mali,	ECA UNCTAD

Project	Countries covered	Executing agency
	Niger, Nigeria, Senegal, Sierra Leone, Togo, Upper Volta	
Mano River Union assistance to the Secretariat	Liberia, Sierra Leone	UNCTAD
Adviser in sales promotion assigned to Africa Groundnut Council	The Gambia, Mali, Niger, Nigeria, Senegal, Sudan	UNCTAD
Sahel groundwater assistance	Chad, Mali, Mauritania, Niger, Senegal, Upper Volta	UNOTC
Study of the hydroelectric and irrigation potentials in the Mano River basin	Liberia, Sierra Leone	UNOTC
Hydrological and topographical study of the Gambia River basin	The Gambia, Senegal	UNOTC
Institutional support for Secretariat of Lake Chad Basin Commission	Cameroun, Chad, Niger, Nigeria	UNDP
Institutional support for OMVS	Mali, Mauritania, Senegal	UNDP
Institutional support for Liptako-Gourma Authority	Mali, Niger, Upper Volta	UNDP
Integrated development of Senegal River basin	Mali, Mauritania, Senegal	UNDP
Predevelopment report on water resources of Chad basin	Cameroun, Chad, Niger, Nigeria	UNDP
Study of optimum supply of electrical energy needs of *Communauté Electrique du Benin*	Benin, Togo	UNDP
Updating of prefeasibility studies of the Mano River	Benin, Togo	UNDP
Indicative development plan for Niger River	Benin, Cameroun, Chad, Ivory Coast, Mali, Niger, Nigeria, Upper Volta	ECA
Niamey UNDAT	Benin, Ghana, Ivory Coast, Niger, Nigeria, Togo, Upper Volta	ECA
Assistance to *Banque Ouest-Africaine de Development*	Benin, Ivory Coast, Niger, Senegal, Togo, Upper Volta	IBRD

Source: J.P. Renninger, *Multinational Cooperation for Development in West Africa.*

What external forces? 161

and the rest of the world, and (vii) projects supporting multinational co-operation in the field of industry.[3]

Strict adherence to the these priorities is evident in Table VII.1, which also shows clearly that the organisations in the UN system have played an important role in the promotion of economic integration in West Africa. Unquestionably, the part played by ECA in the formation of ECOWAS was a crucial one and its continued support along with other arms of the UN will continue to be needed for the future survival of the new organisation.

1(*b*). *The multinational firms* The other area of interest as well as concern as regards integration in West Africa has to do with the place of multinationals. We noted in the last chapter that the displacement and replacement of foreign transnational firms by indigenous companies or the regulation of the activities of the latter within the framework of ECOWAS would be one way of retaining the gains of integration within the region. However, given that foreign multinationals have easier access to capital markets and international financial institutions, advanced technology and capacity to expand production to take advantage of economies of scale, indigenous companies are not in a position to compete effectively with them. Thus the competitive strength of indigenous companies will depend on the degree of statutory protection they receive from the state. Retaining the fruits of integration within the integrated area through state support can take the form of direct financial support to indigenous private firms, reserving investments in certain sectors to these firms only, joint financing of large-scale or common projects (e.g. the Guinea-Nigeria-Liberia joint investment in iron mining in Guinea), joint financing of expensive feasibility studies, and joint development of basic infrastructure of common interest, such as the West African coastal highway which will pass through all the coastal states of the sub-region. Through these means members of a grouping can influence the pattern of intra-union industrial location.

Another way of influencing the distribution of investments within a customs union is to adopt a common investment scheme or at least a 'rationalised' and/or a harmonised arrangement. At present individual West African countries offer a variety of fiscal incentives to foreign investors, including depreciation allowances, tax remission, refunds of duty on imported raw materials, and refund of duty on the imported content of exported manufactures. It is necessary to harmonise the machinery for investment incentive legislation

3. *Ibid.*

throughout the sub-region to ensure complementarity of policies and to discourage abuse by multinational foreign investors. A clear-cut and well articulated system of investment incentives will not only go a long way to attract investments from genuine foreign investors, but will also minimise unnecessary and wasteful competition and uneven distribution of integration-induced industries in the region.

A further instrument for industrial planning by administrative means rather than through the market mechanism can take the form of industrial licensing. Under an industrial licensing system, the main purpose of which in this case is to ensure a fairer distribution of new industries serving the regional market, authorised products could be manufactured only under licence from the union's secretariat. Before the licence can be granted, the firm (whether foreign or local) will agree to abide by a code of behaviour designed to guarantee union members optimum benefits from the investment. ECOWAS, for example, has established a regional company law and investment regulations designed to achieve this purpose.

At its meeting in Dakar on 26 November 1979, the ECOWAS Council of Ministers considered and endorsed a memorandum which provided for the preparation of a 'legal framework for the definition, status and operation of regional industries' and 'a legal regime governing community enterprises'. Consequently, a supranational investment code was prepared, which will be applicable only to undertakings enjoying ECOWAS status. The revised version of this document, which was presented as a Draft ECOWAS Protocol Relating to Community Enterprises, was discussed at the Council of Ministers' meeting at Freetown, Sierra Leone, in May 1981.

Article 3 specifies the conditions for approval of community enterprises, while Article 4 of the Protocol offers investment guarantees and benefits. To prospective investors, foreign and local, these two articles are of the utmost interest. An important role is assigned to integration (Community) industries in the ECOWAS overall strategy. They are expected to promote regional interaction between the member-states, serve as a stabilising element, and contribute to the development of priority sectors. Therefore, the success or failure of industrial development in West Africa within the framework of ECOWAS depends largely on the attractiveness of these provisions to investors and on the co-operation of member-states. The two vital articles read as follows:[4]

---

4. ECOWAS, Draft ECOWAS Protocol Relating to Community Enterprises, ECW/1ANC/III/2 Rev. 2, Lago, 1981.

ARTICLE 3
CONDITIONS FOR APPROVAL OF COMMUNITY ENTERPRISES

1. No enterprise may be approved as a Community Enterprise unless it satisfies the following conditions:
 (a) that no less than 51 per cent of the equity holding of the enterprise is at any time vested in citizens, entities or governments or their agencies of two or more of the Member-States;
 (b) that the majority of its Board of Directors or equivalent body are citizens of the Member-States;
 (c) that its operations affect two or more of the Member-States;
 (d) that its capital in local currency is not less than equivalent of 200,000 units of account and where the investment is to be made in an industrially less developed region of the Community, not less than 150,000 units of account; provided that the Council may prescribe some other level of capital having regard to the scope and nature of the enterprise;
 (e) that its purposes will promote the development policies and programmes of the Community.
2. In addition to the conditions prescribed in paragraph 1 of this Article, an enterprise in respect of which an application is made for approval under this Protocol, shall first have been incorporated as a Société Anonyme, a Public Company or Sociedade Anonime in a Member-State and shall also provide in its articles of association of equivalent document that:
 (a) its headquarters be in a Member-State;
 (b) all its shares shall be registered and shall enjoy the same rights;
 (c) no dealings in its shares shall take place without the approval of the Board of Directors or equivalent body and in no case shall such dealings alter the amount of equity holding reserved for persons, entities or governments of the Member-States as prescribed in sub-paragraph (a) of paragraph 1 of this Article and that all approved dealings in its shares shall be notified to the Executive Secretary;
 (d) all shareholders shall be enabled to exercise in a reasonable manner (particularly those shareholders resident in the Member-States other than the Member-State where the headquarters of the enterprise is) all their rights, particularly with respect to their attendance at meetings of the organs of the enterprise and voting there and of their being kept properly informed of the operations, decisions and accounts of the enterprise;
 (e) voting in respect of important decisions relating to the enterprise such as the alteration of the instrument of incorporation, increase in the capital and dissolution of the enterprise, appointment and removal of members of the Board of Directors or equivalent body, change in the location of the headquarters of the enterprise shall be by a majority vote of two-thirds of the total shareholding;
 (f) the appointment and removal of members of the Board of Directors or equivalent body shall be taken during a meeting of shareholders;
 (g) no change in the composition of the Board of Directors or equivalent

body and in the control of the enterprise shall be permitted which reduces the number and effectiveness of citizens of Member-States on the Board of Directors or equivalent body or with respect to the control of the enterprise, and any change in the composition of the Board of Directors or equivalent body of the enterprise or in its control shall be registered with the Executive Secretariat.

3. Approval for an enterprise under this Protocol shall have regard to the ability of the enterprise to contribute to the following objectives:

(*a*) the development of the Community in general, and in particular the industrially less advanced regions of the Community;

(*b*) the efficient use of the resources of the Member-States and their economic potential;

(*c*) a minimum of 35 per cent value added in the process of manufacture;

(*d*) the expansion and creation of employment within the Community;

(*e*) improved access to the international capital markets;

(*f*) the provision of satisfactory arrangements for the training of citizens of the Member-States in administrative, technical, managerial and other skills with a view to securing the benefit of their knowledge and experience in the conduct of the enterprise;

(*g*) the transfer and adaptation of technology and technical skills to, and absorption thereof by, persons who are citizens of the Member-States;

(*h*) the efficient saving on imports from third countries and the increase of trade within the Community and of exports to third countries that would assist in strengthening the balance of payments position of the Member-States;

(*i*) where the nature of its activities require, the provision of sufficient and adequate environmental and pollution controls and, where the activities are likely to be of limited duration, the restoration of the environment to its original state or as near thereto as circumstances may permit on the cessation of such activities.

ARTICLE 4

INVESTMENT GUARANTEES AND BENEFITS

1. No enterprise approved under the provisions of this Protocol shall be expropriated by the Government of any Member-State except for valid reasons of public interest, whereupon fair and adequate compensation shall be promptly paid.

2. Subject to the provisions of this Article, no person who owns, whether wholly or in part, the capital of an enterprise approved under the provisions of this Protocol shall be compelled by law, during the period the enterprise continues to enjoy benefits under this Protocol, to cede his interest in the capital.

3. Benefits granted to a Community enterprise at the time of approval shall not, except as provided under Article 21 of this Protocol, be altered subsequently to its disadvantage.

4. Where, however, in exceptional circumstances an approved enterprise is taken over or the capital in it is by law ceded to another in the public interest,

What external forces?

the Government concerned shall pay fair and adequate compensation for the take-over or cession.

5. Where there is a dispute as to the amount of compensation payable under this Article the matter shall where applicable be dealt with in the manner specified in Article 23 of this Protocol, and for that purpose any application for the approval of an enterprise under this Protocol shall state that the investor shall consent to the settlement of a dispute in the manner specified, and an approval by the Community shall constitute the consent of the Community to submit to the form of settlement.

6. Community enterprises shall have full legal personality in each of the Member-States and shall enjoy the widest rights and powers as are conferred on enterprises incorporated in such Member-State under its laws.

7. Community enterprises shall enjoy in the Member-States the most favourable treatment with regard to industrial, financial and other incentives or advantages; they may also enjoy the right to operate in sectors reserved exclusively for national enterprises in the Member-States.

8. (*a*) There shall be no restriction with respect to a Community enterprise on:

  (i) the remittance of funds for payment in respect of normal commercial transactions;
  (ii) the remittance of capital, including appreciation, to the country of origin of a Community enterprise in the event of sale or the liquidation of the Community enterprise;
  (iii) the transfer of profits out of the country in which they have been earned after adequate provision has been made for maintenance and replacement of assets and after payment of any tax due in respect of the Community enterprise; the appropriate percentage of transfer shall be determined in the contract of establishment;
  (iv) the transfer of payment in respect of principal, interest and other financial charges where a loan has been granted to a Community enterprise by a non-resident for its purposes in accordance with the approved conditions of the loan;
  (v) the transfer of fees and charges incurred by a Community enterprise in the ordinary course of business outside the country of its principal place of business;
  (vi) the transfer of funds for the payment of holdings in the capital of a Community enterprise;
  (vii) the entry into a Member-State of the requisite foreign managerial and technical personnel for employment or engagement in a Community Enterprise, if the requisite skills are not available in the Member-State.

(*b*) Reasonable facilities shall be provided by the monetary authorities of the Member-States concerned to personnel employed or engaged in a Community enterprise for making remittances abroad in respect of maintenance of their families and other contractual obligations such as insurance premiums and all contributions to provident and pension funds.

9. Further benefits that may be granted a Community enterprise where applicable include:

(a) tariff and non-tariff protection for such duration as may be determined by the Council;
(b) free movement of factors of production relating to the investment within the Community;
(c) the free convertibility of funds among the currencies of the Member-States and freedom from exchange controls and other restrictions on the movement of funds within the Community;
(d) drawback of all import duties paid in respect of goods exported from the Community;
(e) exemption for a period not exceeding ten years as may be determined by the Council from all customs or other import duties on the importation of any plant, machinery and equipment necessary to establish the investment project, in respect of importation of any raw materials and semi-processed goods required in the production process;
(f) the right to carry trading losses incurred in any given year in the first seven years after the commencement of commercial activities forward or back against profits made in other years.
(g) assistance in obtaining land or industrial sites;
(h) the unrestricted grant of all import and export licences, work and residence permits, entry and exit visas and all other legal or administrative authorisations necessary for its purposes;
(i) financial grants in respect of new investments or re-investments as provided for in Article 19 of this Protocol;
(j) any other advantages reasonably incidental to the enjoyment of any of the preceding advantages.

10. An investment contract may, without prejudice to the right of products which otherwise enjoy Community tariff treatment, provide that in respect of a Community enterprise which is engaged in an industrial or economic sector as may from time to time be designated by the Council as being a priority sector, and which introduces a new industrial or economic activity into the Community without causing undue distortion to the economic equilibrium of the Community, no other investment contract under this Protocol may be issued with respect to the same or similar industrial or economic activity for a specified period or with respect to a region of the Community as may be determined by the Council.

11. Where a Community enterprise enjoys the benefits provided for in paragraph 10 of this Article:
(a) the products of that enterprise shall notwithstanding any provisions of the Treaty to the contrary not be subject to any form of tariff or non-tariff restrictions or barriers or to the application of the provisions of Article 26 of the Treaty;
(b) no products which are the same or similar to the products of that enterprise may be imported into the Community except temporarily unless the product of that enterprise and any other enterprises producing the same or similar products within the Community are insufficient in quantity or quality to meet the demand for that product or similar product and where so authorised by the Council.

12. Notwithstanding the provisions of this Article the Council may accord

to a Community enterprise such other benefits, incentives or advantages, other than those set out in this Article, as it may deem necessary or desirable.

With the application of the above measures, the unchecked exploitation of the opportunities created by integration movements in West Africa by multinationals will be minimised. Besides, harmonised treatment of foreign investment in relation to capital and technology transfers will not only help the integration movements concerned but will also aid the cause of the new international economic order.

2. Policy guidelines

The history of economic integration in West Africa should be critically studied as a guide for future action. Weaknesses identified in the defunct schemes should be closely looked at with a view to advising the governments of member-countries to avoid them in the future. Some of the more important problems are briefly touched on here.

— Greater emphasis needs to be placed on the establishment of well-staffed and efficient institutions to run the schemes. Effectiveness and efficiency, initiative and innovation should be the major criteria for defining the scope and authority of the community's organs, the quality of its staff, the terms and conditions of their appointment, and the process by which they are selected. A potential for friction, leading to paralysis, usually exists where the relationship between the Secretariat and other organs of a grouping is not firmly defined. Consider the case of ECOWAS. The Treaty did not specify in detail the relationship between the Secretariat and the Fund for Co-operation, Compensation and Development. Consequently, the Managing Director of the Fund, Dr Romeo Horton of Liberia, refused to take orders from the Secretary-General and chief executive of ECOWAS, Dr A.D. Ouattara. The matter reached crisis point and the Authority of the Heads of State and Government of ECOWAS had to intervene. In the end Dr Horton was replaced by another Liberian, Robert Tubman. Institutional problems of this nature should be avoided through detailed specifications in the Treaty.

— One of the key problems facing all integration schemes is that of unfair distribution of the benefits and costs of integration. There is a clear preference for the equitable distribution of the net benefits of integration, and indeed the survival of a scheme largely depends on this. As for the concept of equity, a useful starting point must be that an integration arrangement will generate net benefits both to the group as a whole and to each individual member. This implies that each member of the group must be at least as well off as a member as if it had not joined the scheme. It is with the distribution of these net benefits after this condition is satisfied that equity and fairness are involved. Since we are concerned with the equitable distribution of the net benefits resulting from integration (i.e total benefits minus the benefits which would have accrued to each member-state outside the integration scheme), it seems difficult to postulate an objective criterion for determining the equity or fairness of any arrangements and their results. What seems compelling, if any major negotiation on integration is to be fruitful, is that there should be substantial agreement from the start among the participants on the general basis for judging the acceptability of an arrangement. As discussed earlier, the arrangement itself should incorporate, among other things, a compensation fund, an industrial licensing system and a common investment policy. However, the guidelines which are accepted will contain some 'political' and 'non-political' constraints which may become necessary in the light of the development objectives and strategy of the participants in order to satisfy the aspirations of member-states — individually and collectively.

— There is an urgent need for the harmonisation of industrial development policies, and machinery with this object would be easier to operate at the planning stage of the development programmes of West African countries. Since the pattern of industrial development largely determines the direction and volume of trade, success in this area will certainly increase the volume of intra-West African trade. Lack of industrial harmonisation and rationalisation, on the other hand, leads to multiplication of identical plants which in turn leads to excess capacity and limited intra-union trade.

— The current payments arrangements and exchange control regulations prevalent in the area are not conducive to the expansion of zonal trade. Although the West African Clearing House discussed earlier exists, it can only be regarded as an interim measure in view of its inadequacies. In the long run the idea of a supra-national West African currency (distinct from national currencies) may have to be considered. Different national exchange control regulations will have to be harmonised within a regional framework. At least, in order to foster intra-regional trade, a special dispensation for relaxation of exchange control regulations applicable to intra-regional transactions and movement of persons may be expedient.

— Another important issue is the pricing policy of regional enterprises. Since integration industries (existing or prospective) cater both for their national consumers and for extra-national ones, the formulation of an intra-union pricing policy is crucial to the distribution of the costs and benefits of integration. In this exercise the unit costs of production of such products, including a normal profit, may not necessarily be indicative of their selling price, although of course the c.i.f. price of competing goods from the rest of the world will set an upper limit to the prices which will be charged after allowing for any tariff differentials. The unit costs of distribution of regional plants will evidently be an important determinant of the distribution of costs and benefits of agreed specialisation.

Furthermore, the price policy adopted will determine the tariff preferences, if any, which may be required to implement the specialisation arrangements for future projects on the trading side. These preferences, in turn, will provide one indicator of the direct costs which the arrangements impose on the participating countries. Each country will have to consider its costs in conjunction with any benefits it derives from its ability to purchase other products at favourable prices, and the net balance after taking this into account will in turn have to be looked at in the light of the costs and benefits associated with the investment in the industries which are allocated to it.

— Like the industrial sector, co-operation in the field of agriculture is of prime importance. Harmonization of agricultural policies should be vigorously pursued in specific areas such as the fixing of producer prices for primary products, the development of large river basins, the regulation of river flows for irrigation, livestock development and marketing, the establishment of an international research station for the Sahel and the creation of sub-regional food reserves.

— Co-operation in the area of transportation is also imperative. Different statistical procedures in use in different countries need to be compiled, defined and harmonised for common use. All means of transport — by road, rail, air and water — should be co-ordinated, restructured and developed to serve the needs of intra-regional trade. The operation of non-economic individual national airlines is against the spirit of integration. The time has come to start thinking of some form of loose merger to ensure economic viability and efficiency.

— International migration has a long history in West Africa partly because of the artificial nature of the political boundaries and partly because of the commercial orientation of the people. Recognising this, Article 27 of the ECOWAS Treaty provides for freedom of movement and residence of persons within the community. The first step in implementing this provision was taken with the signing on 28 May 1979 of the Dakar Agreement, whereby West African citizens are not required to obtain visas before travelling to any country in West Africa, and they can stay without special documents for up to ninety days. A longer stay requires additional residence qualifications. Up to the time of writing, nine countries have ratified the agreement, and thus its operation is restricted to those countries only.

The problem of free movement of persons in the region is complicated, as noted earlier, by the co-existence of unequal partners (i.e strong and weak economies) and the presence of a large number of non-West Africans, especially the Lebanese communities of traders and businessmen who are eager to take advantage of this agreement to tighten their grip on the region's economy. In Ivory Coast, for example, the estimated 30,000 Lebanese inhabitants possess 60 per cent of real estate property, 61 per cent of petrol stations, 32 per cent of travel agencies, 25 per cent of grocery stores, and account for the sale of 83 per cent of shoes and 66 per cent of textiles. This commercial power has frustrated Ivorian businessmen, who are putting pressure on the government to decrease the numbers of the Lebanese community in their country.[5] It is unlikely that other West African countries will open the floodgate for such groups, and this is reflected in the caution with which West African countries generally have so far approached the issue of freedom of movement and residence in the region. Nevertheless, the policy is consistent with the spirit of integration and should be pursued to its logical conclusion provided the necessary checks and balances are instituted.

---

5. Recent political unrest in several West African states has made long-established Lebanese communities of traders and businessmen jittery. Estimated at some 70,000, they are to be found in most West African capitals and they usually control a fair segment of the host-countries' commercial activities and have excellent contacts in most administrations. Because of their relatively dominant position in the commercial sector, radical governments have often found them a convenient lever for consolidating a power base among indigenous businessmen eager for a greater share of the economic pie by blaming them for post-independence economic shortcomings.

VIII
CONCLUSION: PROSPECTS FOR GROWTH

In recent times an unusual bunching of unfortunate events, both external and internal, has had an adverse effect on the development process in West Africa. We shall conclude this study by highlighting and analysing the major factors that have tended to thwart trade-induced growth in the sub-region.

1. *Growth in the 1970s*

The first United Nations Development Decade (1960–70) was a period of socio-economic reconstruction and consolidation for West Africa after the attainment of political independence. The establishment of economic and social institutions and infrastructures conducive to economic modernization, the deliberate integration of national economies and the decolonisation of economic and political goals formed the focus of policy objectives. It could therefore be argued that a large portion of investments made during the 1960s went into long-term non-self-liquidating projects whose impact on growth could not be felt in the short run. In other terms, it was expected that the contributions of these projects would spill over to the 1970s, when total growth, given the expected increase in investments of the self-liquidating variety, would be appreciably higher. Unfortunately, the record of the 1970s bears no eloquent testimony to this expectation.

As can be gleaned from Table VIII.1, most of the countries of West Africa maintained a higher average growth rate of the GDP in the 1960s than in the 1970s. The only notable exceptions are Benin, Mali, Niger and Nigeria. Guinea seems to represent a static case, but Benin, Mali and Niger undoubtedly ended the 1960s with a relatively low economic base which tended to exaggerate marginal improvements in economic performance in the 1970s, while the case of Nigeria can be largely attributed to the contributions of the oil sector.

Thus, during the 1970s, those internal factors that hampered growth and trade expansion in the 1960s still persisted, in some cases with increased vigour. These include political fragility, institutions ill-suited to actual needs, a climate and geography hostile to development, an over-extended public sector *vis-à-vis* the available administrative capacity, neglect of export-oriented industries, and a bias against agriculture. Since independence all West African countries except Guinea, Cape Verde, Ivory Coast and Senegal have experienced political and military turmoil of one kind or another, leading ultimately to a change of government. Given the existence of diverse cultures and languages in individual countries of the region, the process of national integration (i.e. building new institutions and loyalties) inevitably involved strife. This in turn resulted in political instability which invariably affected the process of socio-economic development in several negative ways. It forced post-independence leadership to give high priority to short-term political objectives; it triggered large-scale displacement of people and induced a diversion of resources to military spending. For instance, the share of GNP devoted to military purposes by the smaller countries of West Africa averaged around 4 per cent in the 1970s, while the comparable figure for the 1960s was only a little over 2 per cent.[1] Surely, military spending can hardly be regarded as a form of investment in directly productive activity.

Another factor that constrained trade-induced growth over the

Table VIII.1
WEST AFRICA: GROWTH OF GDP, 1960–79

	Population (million) mid-1979	GNP per capita (US$) 1979	GDP (average) rate (%)	
			1960–70	1970–79
Benin	3.4	250	2.6	3.3
Cape Verde	0.3	260	n.a.	n.a.
The Gambia	0.6	250	5.4	2.8
Ghana	11.3	400	2.1	– 0.1
Guinea	5.3	280	3.5	3.6
Guinea-Bissau	0.8	170	n.a.	n.a.
Ivory Coast	8.2	1,040	8.0	6.7
Liberia	1.8	500	5.1	1.8
Mali	6.8	140	3.3	5.0
Mauritania	1.6	320	n.a.	1.8
Niger	5.2	270	2.9	3.7
Nigeria	82.6	670	3.1	7.5
Senegal	5.5	430	2.5	2.5
Sierra Leone	3.4	250	4.3	1.6
Togo	2.4	350	8.5	3.6
Upper Volta	5.6	180	3.0	– 0.1
Total	144.8	500 (av.)	n.a.	5.9 (av.)

Source: World Bank and UNCTAD data.

1. World Bank, *Accelerated Development in Sub-Saharan Africa: an Agenda for Action*, Washington DC, August 1980, 11.

period under consideration had to do with institutional adaptation. The institutional heritage of post-colonial West Africa needed to be adapted to the new political realities in order to meet national needs and aspirations. Governments of the region inherited many subregional organisations such as the West African Cocoa Research Institute, the West African Institute for Oil Palm Research, the West African Currency Board, and the West African Examinations Council and the Federation of French West Africa, to name a few. Most of these proved unsuitable to the new national realities and were either disbanded or reorganised; and at the national level the systems of local government, civil service, education and the like bequeathed by the colonial administration needed restructuring. But the reorganisation and restructuring of these institutions to make them appropriate for today's needs — at both the national and regional levels — have proved costly and difficult; and, in some areas, the search for the most suitable systems continues.

There is also the question of climate and geography. Unquestionably, some of the poor performance of the 1970s was due to bad weather. The Sahel region of West Africa experienced a quick succession of drought years between the late 1960s and 1972–4, with only one or two years of recovery in between. A period of satisfactory weather in the mid-1970s was then followed by a number of poor years, starting in 1977–8. This resulted in a sharp drop in crop production and severe losses of livestock across the Sahel. Geography, too, has had an impact. The topography and pattern of population concentrations create special transport needs and problems. The inadequacy of the existing network of roads and the dire need for integrated regional transport and communications infrastructures have long been appreciated.

Recognising the role of effective transport system in the expansion of West African trade, and the parallel need for effective communications facilities, the United Nations General Assembly (in its Resolution 32/160 of 19 December 1977) proclaimed a Transport and Communications Decade in Africa (UNTACDA) for the years 1978–88. The principal goal of the Decade at the regional level is to integrate the national and regional transport and communications infrastructures with a view to increasing their effectiveness and to foster intra-West African trade.

The other issue turns on the over-extension of the public sector. After independence West African states inherited unevenly developed economies with rudimentary infrastructures. Markets often functioned imperfectly and foreigners dominated trade and most modern businesses. To capture the commanding heights of their economies and speed up development, post-independence governments expanded the public sector by moving into commercial and productive activities previously reserved for the private sector. Conceptually this was a sound policy option but its implementation left much to be desired. It is now widely evident that most public enterprises in the region have been found to be corrupt and inefficient, and have needed to be heavily subsidised by the taxpayer. To the extent that certain public enterprises — like airways, state trading companies, manufacturing enterprises, service and supply agencies — constitute a drain on their economies, growth has inevitably been slower than might have been the case with available resources. And this accounts in part for the comparatively poor record of the 1970s.

Improved performance by public agencies seems to be a *sine qua non* for the attainment of accelerated growth. The organisation and management of economic activities within the region require urgent review to determine how the resources and energies of all economic agents can be better mobilised for development. Governments now accept the need to reorganise public policy-making institutions and procedures to make them responsible and efficient.

There is also the problem of neglect or inadequate promotion of export industries. The main cause of rising current account deficits and shortages of foreign exchange in the 1970s was not the terms of trade *per se* but the slow growth of exports and the acceleration of imports. This is unmistakably demonstrated in Tables VIII.2 and VIII.3. Of all the thirteen countries of West Africa for which data are readily available only two countries (Mali and Niger) experienced an increase in their volume of exports during the 1970s as compared with the previous decade. For the rest there was either only a marginal increase or a decrease. Table VIII.2 shows that the average annual growth rate in volume of exports was negative in seven countries throughout the period, with Benin, Ghana, Sierra Leone and Togo being the worst sufferers. Over the same period a fairly high level of imports was maintained in these countries, particularly Nigeria, Ivory Coast and Togo. Only Sierra Leone registered a decrease in imports over the period. It is therefore not surprising that current account deficit as a percentage of GDP in oil-importing African countries grew from 2.4 per cent in 1970 to 9.5 per cent in 1975, declining slightly to 9.2 per cent in 1980.[2]

That the decline in aggregate volume of exports is the principal source of this problem can be further attested by Table VIII.3, which clearly shows that the terms of trade for all the countries of the region in the 1970s were not notably worse that those of the 1960s.

---

2. See *Ibid.*, 19.

Table VIII.2
GROWTH OF MERCHANDISE TRADE

	Merchandise trade (US$ millions)		Average annual growth rate (Volume, %)			
	Exports 1979	Imports 1979	Exports 1960–70	Exports 1970–9	Imports 1960–70	Imports 1970–9
Benin	190	357	5	«11.4	7.4	6.3
Ghana	1,096	993	0.2	«7.2	«1.5	0.1
Guinea	373	347	n.a.	n.a.	n.a.	n.a.
Ivory Coast	2,515	2,491	8.8	5.8	9.7	10.1
Liberia	506	487	18.4	2.3	2.9	2.3
Mali	177	180	3.0	6.7	«0.4	5.5
Mauritania	147	257	50.7	«1.1	4.5	5.5
Niger	n.a.	n.a.	6.0	11.7	11.9	6.5
Nigeria	18,073	12,399	6.6	«0.3	1.6	20.6
Senegal	421	756	1.2	«0.8	2.3	4.5
Sierra Leone	205	297	0.3	«6.5	1.9	«3.0
Togo	251	441	10.5	«2.5	8.6	9.8
Upper Volta	81	254	15.9	3.1	8.5	5.2

Source: World Bank data.

Using 1975 as the base year, only Liberia and Mauritania suffered marginal deterioration in their terms of trade, while the rest experienced substantial improvements in the 1970s compared with the preceding decade. This was particularly the case in Nigeria, Senegal and Togo. Thus, even allowing for the statistical imperfections of trade figures, the picture which emerges from the foregoing is one of near-stagnant or declining export volumes for the region as a whole during the 1970s. At the same time, a high level of imports was more or less maintained with a consequent worsening of the region's balance of trade and payments.

Indeed, for the continent as a whole its share of non-fuel world trade fell markedly during the 1970s. Its share of non-fuel world exports fell from 2.4 per cent in 1970 to 1.2 per cent in 1978; and its share of developing country non-fuel exports also dropped from 18.6 per cent in 1970 to 9.2 per cent in 1978. The reason for this poor record is essentially structural. Africa is more dependent than any other region on exports of primary products. For example, thirty-two major resource commodities accounted for about 70 per cent of its non-fuel exports during 1976–8, compared with 35 per cent for all developing countries and 10 per cent for the world. Since world trade in most primary products grows more slowly than world trade in manufactures, Africa's share of total trade would, *ipso facto*, tend to fall.

Table VIII.3
WEST AFRICA: TERMS OF TRADE, 1960-79

	1960	1965	1970	1971	1972	1973	1974	1975	1976	1977	1978	1979
Benin	114	128	129	137	116	119	126	100	116	115	100	97
Gambia	106	107	109	119	125	102	99	100	93	116	113	93
Ghana	111	76	121	90	80	85	96	100	94	152	197	144
Ivory Coast	113	99	127	113	104	103	104	100	112	165	155	129
Liberia	255	143	131	131	124	110	102	100	103	103	94	88
Mali	107	118	117	119	124	115	100	100	122	119	110	95
Mauritania	149	149	133	130	119	103	92	100	104	94	80	78
Niger	98	114	109	114	128	122	113	100	96	102	102	90
Nigeria	32	28	32	34	34	39	103	100	105	110	100	119
Senegal	71	70	79	84	84	73	93	100	94	103	94	76
Sierra Leone	121	117	136	134	112	108	103	100	104	121	115	108
Togo	56	49	59	51	46	48	91	100	88	106	103	82
Upper Volta	88	124	117	118	125	115	108	100	109	109	105	94

Source: UNCTAD, *Handbook of International Trade and Development Statistics*, 1980.

Associated with the problem of slow growth of trade in primary products *vis-à-vis* manufactures is the issue of low productivity in the region's agriculture. Agriculture is at the heart of African economies, and most of the population earn their livelihood from it. The transport, industrial processing and trade sectors depend on the production of agricultural commodities, and incomes earned in this sector provide markets for domestically produced goods and services. Increased agricultural output is therefore the single most important determinant of overall economic growth. But with the exception of Ivory Coast, the growth rate of agricultural production in the 1970s was less than the rate of population growth throughout the region (Table VIII.4). Consequently exports stagnated, food production *per capita* fell, commercial imports of food grains rose, and more people shifted their consumption to wheat and rice thereby increasing their food dependency.

There are specific explanations for the poor agricultural record of the last decade. These include disruptions caused by wars and civil strife, drought and poor rainfall patterns during the 1970s, neglect of agriculture by government and development theorists, and misallocation of investment through over-emphasis on large-scale government-operated schemes. Also agricultural and economic policies and institutional frameworks were not conducive to increasing output: official prices were often too low, marketing systems were too uncertain, inefficient and uncompetitive; input supplies were too irregular, and the participation of farmers in rural

Table VIII.4
WEST AFRICA: GROWTH RATES OF AGRICULTURAL
PRODUCTION, 1969/71–1977/9.
Average annual growth rate in volume (%)

4@	3–4	2–3	1–2	0–1	⅔0
	Ivory Coast	Benin Liberia Upper Volta	Guinea-Bissau Mali Niger Nigeria Senegal Sierra Leone	Gambia Guinea	Ghana Mauritania Togo

Source: FAO, *Production Yearbook* tapes.

affairs was too limited. The agricultural extension effort was weakened by unfavourable policies, deficient research output, and the organisational deficiencies of the public sector agencies which were responsible for spearheading rural development. Because of its critical role, growth-oriented policies are urgently needed in this sector.

2. Growth through trade

World trade grew by an average of 5.7 per cent a year in the 1970s, after almost 8 per cent a year in the 1960s. Despite this slowdown of the growth of total trade, the non-fuel exports of developing countries grew faster — over 7 per cent a year in the 1970s, compared with 5 per cent in the 1960s. On the face of this evidence, one would have expected a corresponding level of trade-induced growth and industrialisation but that did not happen — at least, not for all developing countries.

To start with, less-developed countries (LDCs) are a mixed bag, embracing the oil-exporters and the non-oil exporters which are also commonly classified as middle- and low-income countries respectively. The two groups faced totally different experiences during the 1970s because the most striking changes in the pattern of world trade in that decade have resulted from the increase of fuel prices. World trade in fuels increased from US$29 billion in 1970 to $535 billion in 1980, or from 7 to 21 per cent of world trade.[3]

The low-income developing countries benefited very little from this phenomenal growth. The overwhelming majority among them resorted to an inward-oriented trade policy. They reduced their current account deficits by curbing imports (and hence growth)

---

3. World Bank, *World Development Report*, Washington DC, 1981, 20.

rather than by expanding exports. The middle-income countries (some of them oil exporters) on the other hand were able to take advantage of the trade expansion: oil exporters increased their sales and increased their imports while other middle-income countries adopted a more outward-oriented trade policy and expanded exports, so reducing their current account deficits to levels financeable in the medium term, without sacrificing growth.

Unfortunately, the vast majority of West African countries did not experience the growth of trade and economic expansion which many of the middle-income countries witnessed in the 1970s. Apart from Ivory Coast, Nigeria, Ghana, Senegal and Liberia, all the (eleven) remaining countries of West Africa[4] are classified by the World Bank as belonging to the thirty-six low-income developing countries of the world with a *per capita* income of $370 or less in 1979. This group fared badly in the 1970s.

The weakness of the low-income countries' primary export price — as indicated earlier — reflects both their concentration in commodities for which demand is sluggish and their inability to vary their output mix as relative prices change. For example, the annual price fluctuation of thirty-three non-oil commodities of export interest to West Africa averaged 5 per cent a year in the 1950s and 1960s but had increased to 12 per cent in the 1970s. This erratic trend defies arrest, if only because a permanent and effective commodity agreement for every primary export product would be all but impossible.

Another related inflexibility issue holding back the low-income countries relates to the level of processing of their raw materials, in contrast with what is happening in many middle-income countries. Undoubtedly, tariff and non-tariff barriers against processed products are still an obstacle to increased processing for exports but a general lack of industrial skills and capacity may be equally important. But until their exports deepen into processed materials and more sophisticated manufactures, their export prices and earnings will surely continue to be closely tied to random movements of international demand.

Of course, over the longer term, their trade prospects would be primarily a question of their own policies. The export-GDP ratio of the low-income oil-importers of Africa fell from 23 per cent in 1970 to 16 per cent in 1980 (Table VIII.5). Over the same period manufactured exports dropped from 11 to 9 per cent and the import of food

---

4. Eight of these (Benin, Cape Verde, Gambia, Guinea, Guinea-Bissau, Mali, Niger and Upper Volta) are also among the thirty least developed countries of the world according to the UN General Assembly classification.

Table VIII.5
STRUCTURE OF MERCHANDISE TRADE, LOW AND MIDDLE INCOME OIL IMPORTERS, 1970–80 (%)

	Export-GDP ratio	Composition of merchandise exports		Composition of merchandise imports		
		Manufactures	Non-fuel primary	Manufactures	Food	Fuel
1970						
Low-income oil importers						
— Africa	23	11	86	77	11	9
— Asia	7	54	43	64	21	5
Middle-income oil importers	22	33	58	69	12	10
1980						
Low-income oil importers						
— Africa	16	9	80	11	16	31
— Asia	9	47	50	38	14	39
Middle-income oil importers	24	46	36	53	11	28

Source: World Bank, *World Development Report*, 1981, 26.

increased from 11 to 16 per cent of total imports, while oil imports catapulted from 9 to 31 per cent within a decade.

Juxtaposing the African and Asian figures, a different picture emerges (see Table VIII.5). Some Asian countries like India can afford, because of their very size, an inward-oriented development path which enables them to maintain a low export-GDP ratio relative to African countries while at the same time exporting more manufactures in total than African countries. They have also managed to reduce their imports of manufactures and food over the past decade. Obviously, these are indices of development.

Most West African countries cannot afford an inward-oriented development policy in view of the small size of their domestic markets. For small countries with limited human and material resources, an outward-oriented strategy, which necessarily fosters an open economy, has often proved a feasible policy option. Even so, a number of West African countries have an extremely limited base of physical and human resources: and in some others the existing base is actually diminishing — for example, through over-exploitation and erosion of farmland along with emigration of the younger and better-trained workforce.

The problem of emigration of the economically-active population is of course an important issue in countries like Benin, Mali, Niger, Upper Volta and now Ghana. These are the net exporters of labour while the net receivers are Ivory Coast, Nigeria and Senegal and, to a less extent, Sierra Leone and Liberia. It is generally acknowledged that foreign exchange remittances are a major benefit to the emigrant country. And many governments of emigrant countries have attempted to increase total remittances, attract remittances through official channels and encourage workers to put their savings into 'productive investments'. In Benin remittances grew at the annual rate of 26.8 per cent over the period 1967–78, in Mali at 19.6 per cent in 1967–78 and in Upper Volta at 15.6 per cent in 1969–78. Besides, the ratio of remittance inflows to merchandise exports was high in these countries. The percentage of remittances to exports was 59.6 in Upper Volta in 1978–9, 33 in Mali and 16.6 in Benin.[5]

In spite of their foreign earnings and other possible advantages, it is arguable whether on balance these labour-exporting countries really gain, particularly since most of them are in greater need of trained manpower than the labour-receiving states. One is therefore inclined to think that changes in trade policy alone cannot be regarded as a sufficient condition to accelerate the development of

---

5. G. Swamy, *International Migrant Workers' Remittances: Issues and Prospects*, World Bank staff working paper, 481, 1981, 9.

many countries of the region. A frontal attack on poverty of resources, including human capital, must be part of the overall development strategy.

3. Prospects for the 1980s

The low-income countries of West Africa ended the decade of the 1970s in a mood of disillusionment. Some entered the 1980s with a worsening of external conditions and with chaotic domestic policies: in most of them high unemployment co-existed with double-digit inflation, external deficits soared as total exports declined and, worse still, domestic food demand grew faster than total supply. Hasty or ill-considered policies attempting to grapple with these problems often led to more problems. But, provided concerted efforts are made to reverse the situation through the pursuit of appropriate domestic and foreign trade policies, the prospects for the 1980s should be somewhat more favourable for West African oil-importers than the experience of the recent past. At least four areas of policy manoeuvre immediately suggest themselves, namely increased production of exportables, trade promotion strategy, consolidation of regional economic groupings, and increased foreign aid.

3(*a*). *Increased production of exportables.* World trade in the twenty-two non-fuel commodities (i.e. copper, iron ore, bauxite, phosphate, manganese, zinc, tin, lead, coffee, cocoa, sugar, tea, groundnuts, groundnut oil, palm oil, beef, bananas, maize, timber, cotton, tobacco and rubber) of greatest export interest to the region is projected to increase by 2.9 per cent annually during the 1980s as against an annual average increase of 1.5 per cent in the 1970s for the same product group.[6] Also the weighted price of Africa's non-fuel commodity exports is expected to rise slightly so that the average value of world trade in these commodities will show an annual increase of 3.4 per cent. There is no reason why West African countries should not try to step up their domestic production of exportables to take advantage of these encouraging forecasts.

Furthermore, although the future path of oil prices defies precise estimation, it is projected that the price of oil will increase annually by no more than 3 per cent in real terms during the 1980s.[7] At that rate the relative price of oil would increase by slightly over one-third during the 1980s, which is surely small in comparison to the increases in the 1970s, although with the increased weight of oil in total imports, even relatively small price increases will have a large

6. World Bank, *op. cit.*, 22.
7. *Ibid.*, 39.

impact. In fact, for the typical oil-importing African country, a 3 per cent annual increase in the price of oil implies a 0.7 per cent annual decrease in the purchasing power of exports. However, not all countries will be affected: some present-day oil importers like Ghana, Ivory Coast and Cameroun have prospects for developing domestic petroleum resources which will at least satisfy their domestic needs.

3(*b*). *Export promotion strategy*. Since the growth of trade in Africa's main exports is lower than that of overall world trade because of dependence on primary exports, the drive to increase the volume of exports must go hand-in-hand with aggressive trade promotion and diversification. Increased output is one thing but market access is another. Certainly, the key question of the decade is how to achieve easy access to markets (i.e. an open world trading system) in the face of slow growth and mounting protectionist pressures in the industrialised countries.

A set of promising trade negotiations was concluded in 1979 under the aegis of the General Agreement on Tariffs and Trade (GATT). Three major results appear to have been achieved: a substantial reduction in tariffs; a refinement and improvement of international rules on non-tariff measures; and the adoption of a framework of procedural arrangements to encourage and facilitate adherence to the agreements on the part of signatory-countries.

On paper these results look like major achievements but in practice they have dismally failed to allay the fears of LDCs over the currently rising tide towards protectionism. LDCs cannot possibly reap the benefits of their comparative advantage in the face of protectionism. As might be expected, they have expressed keen disappointment with the results on two grounds: first, because the reductions on products of special interest to LDCs fell short of the average cuts;[8] secondly, because the most-favoured-nation (MFN) reductions involved an erosion of the margins of preferences enjoyed by the developing countries under the general system of preferences (GSP). To these should be added two further reasons why the LDCs will continue to face problems in their efforts to export manufactures to advanced countries. Tariffs are of little importance compared with non-tariff barriers to trade such as subsidies and countervailing duties, technical regulations (such as

---

8. Average tariffs on industrial products were reduced by approximately one-third (38 per cent) calculated as a simple average or 33 per cent on an import-weighted basis. But reductions on products of interest to LDCs are smaller than the average cuts — 25 per cent versus 33 per cent on a weighted basis. See I. Frank, *Trade Policy Issues for the Developing Countries in the 1980s*, World Bank staff working paper, 478, 1981, 4.

health, safety and protection, customs valuation, import licensing, quotas etc). Taken together these affect the volume and value of trade more than mere tariff charges. The other reason has to do with the failure of the 1979 Multinational Trade Negotiations (MTN) to agree on a safeguard code defining protective measures to offset the serious harm done to domestic producers from import competition. Because measures have increasingly taken the form of quantitative restrictions (QRs) rather than tariffs, and because the QRs have increasingly been in the form of voluntary export restraints (VERs) rather than formal import quotas, nothing in the field of trade policy is of greater importance to LDCs than the establishment of a more effective international discipline over safeguard measures. Normally, it is hard to see how LDCs can effectively and aggressively sell manufactures in world markets in the absence of an internationally agreed safeguard code.

3(c). *Consolidation of regionalism.* As a consequence of the current gloomy international environment, LDCs and West Africa in particular should face the reality of shrinking opportunities to expand exports to the industrial countries. West African countries, like other LDCs, have indeed been seeking for alternative strategies of economic development. It is from this standpoint that the formation of the Economic Community of West African States (ECOWAS) becomes important.

Economic integration is a hybrid strategy of development. It takes something from both the autarkic and the 'outward-looking' paths of growth. It takes from autarky the protection of the integration area from outside competition through the common external tariff, and from the 'outward-looking' path of development it takes the opening of the national markets of each of the member-countries to regional competition. Thus the integration programmes involving LDCs should be seen much more as an alternative path of economic development than as a way of allowing a better allocation of a given stock of factors of production.[9]

Although ECOWAS requires tremendous structural adaptations and changes to succeed, including strengthening transport and communication links, reducing of monetary and commercial policies that inhibit and distort intra-regional trade, and promoting joint projects in industry, education, research and regional institutions with adequate staff and budgets that could become major

---

9. See Eduardo Lizano, 'Integration of Less Developed Areas and of Areas on Different Levels of Development' in F. Machlup (ed.), *Economic Integration: Worldwide, Regional, Sectoral*, London: Macmillan, 1976, 276.

instruments of co-operation and integration, its formation represents a milestone in the development of intra-regional trade in West Africa. In concrete terms, the first five years of ECOWAS existence — 1976-80 — were spent establishing a functional regional structure and framework and working out and adopting the basic operational policy guidelines and measures necessary for a smooth take-off. The next five-year period — 1981-5 — will be the most trying period for the young organisation. This period will witness the actual implementation of several community plans and projects, which will necessarily require a painful process of adjustment at the national level by way of adapting existing policies, systems and practices to conform to community requirements. Given a strong political will and a conviction that the long-term benefits to be derived individually and collectively will clearly outweigh the short-term losses, members will accept the short-term economic sacrifices. Even so, members have dragged their feet over implementing already agreed measures. By the end of the first half of the second five years of ECOWAS, few concrete results had emerged, particularly in trade liberalisation.

3(*d*). *Aid*. In general Africa needs more aid than any other continent, for reasons which are not far to seek. As we have seen, it has twenty of the thirty least-developed countries of the world, eight of which are in West Africa alone. Growth prospects for African countries are not good because of over-dependence on fluctuating primary products; and because their limited creditworthiness also makes them highly dependent on concessional capital (aid). The net effect of this unhappy situation has been a steady growth in recent years of current account deficit in LDCs. For low-income oil importers, it grew from US$3.6 billion in 1970 to US$9.1 billion in 1980 while that of middle-income oil importers rose even faster from US$14.9 billion in 1970 to US$48.9 billion in 1980.[10] This was bridged principally by aid, commercial loans and depletion of reserves.

However, since the prospects for increased aid, rapid expansion of exports and the sharp reduction of imports are not particularly good, the deficit gaps are expected to widen in future. Also inevitable is the continued growth of debt-service ratios. Table VIII.6 depicts this ugly trend up to the end of the 1980s with debt-service ratios for all oil-importers increasing steadily from 8.4 per cent in 1977 to 19.8 per cent in 1990. Even for oil exporters, a marginal increase is noticeable over the period, except for 1990 for which the estimate shows a slight decline.

Available evidence points to the urgent need for increased aid to

---

10. World Bank, *World Development Report*, *op. cit.*, 49.

Table VIII.6
ACTUAL AND PROJECTED SUB-SAHARAN AFRICAN DEBT-SERVICE RATIOS (%)

Category	1977	1978	1980	1985	1990
Oil Importers	8.4	10.6	15.8	17.6	19.8
Low-income	8.4	10.4	19.2	19.5	19.9
Middle-income	8.1	10.1	13.6	16.2	19.3
Oil Exporters	1.8	3.3	3.5	4.6	4.1

Source: World Bank projections (ADSA, p. 129).

the LDCs in the 1980s particularly the low-income countries of Africa. As the Brandt Commission Report emphasizes, funds for development must be recognised as a responsibility of the whole world community, and placed on a predictable and long-term basis.[11] Given the interdependence of the economies of the developed and developing countries, the developed countries surely do not want to see a further deterioration of the general environment for economic development in Africa. It is now widely recognised that economic development in one part of the world, through its spillover effects, promotes stability and growth in the other parts. The assertion that developing-country growth can directly affect developed-country wellbeing is supported by a report prepared for UNCTAD by economists at the University of Pennsylvania. The report concludes that a 3 per cent increase in the growth-rates of the non-oil-exporting developing countries could result in an annual increase of 1 per cent in the growth-rates of the OECD countries; and that this 1 per cent increase would amount, for the industrialised countries, to the equivalent of about $45 billion plus its job creation and other secondary effects.[12]

To this end, the rich countries should at least try to achieve the existing aid target of 0.7 per cent of their GNP during the present decade. In the words of Lester Pearson, 'international development is a great challenge of our time. Our response to it will show whether we understand the implications of interdependence or whether we prefer to delude ourselves that the poverty and deprivation of the great majority of mankind can be ignored without tragic consequences for all.'[13]

---

11. Willy Brandt (Chairman), *North-South: Programme for Survival*, London: Pan Books, 1980, 52.
12. See *The United States and World Development: Agenda 1979*, New York: Praeger, 1979, 52.
13. L. Pearson (Chairman), *Partners in Development*, London: Pall Mall Press, 1969, 11.

4. Policy recommendations

In this chapter we have tried to show that a set of internal and external factors has inhibited the expansion of West African trade and growth. These factors have persisted since independence sometimes with growing intensity, and their removal would require not just marginal changes in policy and trade relations but profound structural adjustments, both internal and external. In order to arrest the current gloomy situation and reorient the West African economies towards the paths of trade expansion and growth, certain policy actions and initiatives are inescapable.

— Given the interdependence and mutuality of interests between the rich and the poor, the world community through its development agencies should give priority attention to the trade and growth problems of the poverty belts of Africa, especially West Africa.

— West African countries should strive to increase their domestic savings rates, productive investment and overall productivity in the face of fast growing population and urbanisation.

— While it is true that light industries producing durable and non-durable consumer goods have advanced a long way, there is still a considerable margin for import substitution in these sub-sectors in most West African countries. And, in addition to import substitution policy, West African countries with the necessary resources should move into industries producing capital goods and some basic intermediate goods, such as chemicals and metals. This phase of industrialisation would be quite feasible within the framework of ECOWAS.

— Although the internal policies of African countries will be fundamental, a major co-determinant of their prospects will be the extent to which markets in the industrial countries remain open to their exports. The North's intellectual commitment to trade liberalisation is a farce where North-South trade is concerned. The trend has been towards increased protectionism in the guise of protecting jobs in the North. It is therefore recommended that industrial countries should refrain from imposing any additional trade barriers in the future, so as to enable LDCs to reap the benefits of their comparative advantage.

— The volume and real value of international aid must be increased. The rich countries should aspire to achieve the existing target of 0.7 per cent of their GNP by 1985 and, as the Brandt Commission recommends, advance towards 1 per cent before the year 2000.

— Faced with external obstacles to trade expansion, the formation of regional economic groupings becomes a *sine qua non* for African development. The Lagos Plan of Action, which aims at the economic integration of Africa by the year 2000 through regional groupings, takes this position.[14] West Africa has done well to have established ECOWAS, but problems remain. Profound structural changes and adjustments are needed to make integration schemes work. And as ECOWAS is about to enter into its full operational phase, there is a smouldering fear that if the issue of non-compliance with agreements and obligations by some member-states persists it might pose a danger before long. Concerted efforts should be made to remove this source of danger before it takes root.

— The effect of high oil prices on the economies of low-income oil importers of West Africa means that a common solution to the problem must be sought, possibly by way of an exclusive concessionary price over a certain critical minimum level of imports, which will be acceptable to OPEC, donor agencies and the countries concerned.

5. Conclusion

Because a major policy goal in Africa is rapid economic development, and the problem of accelerated development has to be viewed from a wider perspective, regional economic integration should be part of the solution. Economic analysis can provide some guidance to the benefits and costs of a given scheme, but it is incapable of ensuring the success of an existing scheme or of forecast when a particular proposal will get off to a good start. We need to look beyond economics for a fuller answer. In a region that is passing through a historical phase of political instability, nothing is certain and anything is possible. Time alone can give the answer.

We have shown how West Africa progressed from pre-colonial disintegration to colonial integration and from post-independence disintegration to the integrative efforts of recent years. Over the decades many lessons have been learnt and while we have discussed most of them a few can be mentioned again.

The theory of economic integration must be an integral part of the theory of the political economy of underdevelopment. Underdevelopment in a country implies not only an insufficient development of physical and human resources, but also the absence of administrative structures, and expertise, as well as of the tradition of international co-operation, all of which are indispensable for carrying out ambitious economic integration scheme. Furthermore, with few exceptions, the national markets of West African countries are too narrow to sustain medium and large-scale plants needed for an

---

14. OAU, *Lagos Plan of Action for the Economic Development of Africa, 1980–2000*, 1981.

industrialisation take-off; hence economic integration must be seen as a pre-requisite for rapid industrialisation. Also, the acceptance of economic integration as an essential part of the development process involves the surrender of an element of political sovereignty, which remains a bitter bill for many LDCs to swallow.

The history of past and current integrative efforts provides a wealth of instructive experience from which lessons can be drawn for future action. Some policy suggestions have been offered along the lines dictated by the history and circumstances of the region. But we cannot be sure how readily sound economic advice will be accepted and put to use in view of the trade-off which takes place between economic and political considerations. This is particularly serious where political changes occur rapidly and regimes and alignments can alter overnight. Indeed, as noted earlier, political instability can safely be regarded as one of the problems facing the orderly development of economic integration in West Africa.

Nevertheless, this should not give rise to despair. The successful formation of ECOWAS represents a breakthrough in the history of economic integration in West Africa. The progressive implementation of some aspects of its Treaty provisions is a sure sign that the enthusiasm and momentum generated by its formation are still present. Despite the problems discussed in Chapter VI, ECOWAS has recorded a measure of success to date. At least six different Protocols have been signed, including that on Free Movements of Persons and Right of Residence, which has actually been ratified by the required number of member-states and is now effectively in operation. The Secretariat has completed most of the preliminary studies required for the Community's take-off, and the Fund is about to implement its first field project. In the area of inter-state relations, some remarkable progress has been made within the ECOWAS framework. Through the successful diplomatic manoeuvres of three ECOWAS members (Liberia, Gambia and Togo), relations between Guinea and her neighbours Senegal and Ivory Coast have been normalised. Members have also been engaged in joint economic ventures: Nigeria has invested in Guinea's national iron mining company, in which Liberia also has shares. There are similar investments between Nigeria and Niger in the fields of food production, transport and telecommunications, and between Nigeria and Benin. Apart from these concrete achievements, the establishment of the Community itself represents a historical landmark since it was the first *all-embracing* integration scheme ever attempted in West Africa.

The prospects for profitable co-operation within the Community are good, despite the existing problems and difficulties. West

African countries at the present time strictly do not possess a very strong internal dynamic for continuous progress towards effective integration, so deliberate and sustained efforts are called for. Member-states should not allow fears of domination and political and ideological differences to slow down the momentum that has been generated. ECOWAS may have many teething problems and no one expects the successful integration of sixteen different economies to be easy, but it can be done, given a measure of community spirit and strong political will. The sacrifices needed for the benefits of economic integration to be reaped would surely be forthcoming. History is generally on the side of perseverance.

EPILOGUE
ECOWAS POLICY ON POPULATION MOVEMENT AND ITS IMPLICATIONS

1. *General*

Population migration in West Africa is not a new phenomenon. Intra-regional migration began long before Europeans arrived on the West African coast, and increased slowly but steadily up to the beginning of the First World War. There were temporary interruptions during the First and Second World Wars, but the floodgate was then thrown wide open again, and by 1960 population movements had become a matter of serious policy concern to the net labour-importers of the time. For example, by 1960 Ghana had 828,000 immigrants or the equivalent of 12 per cent of its population at the time.

Before independence, the British and French colonial administrators deliberately encouraged free mobility of labour in the region within their respective areas of jurisdiction. This they did for economic and administrative reasons. Thus so long as colonial rule lasted, the individual French West African was easily acceptable anywhere in French West Africa and the same could be said of the British West African. However, with the attainment of political independence and the emergence of national policies, which influenced the performance of individual national economies, the scale of immigration started to worry policy-makers, particularly in the coastal states. Even so, immigrants in the region suffered little disturbance up to the mid-1960s and they enjoyed almost unrestricted entry to the economically strong countries.

The concentration of immigrants in the region has always been dictated by the 'perceived' economic attractiveness and opportunities which the receiving country offers. Up to the late 1960s Ghana was a country of immigration attracting workers mainly from its neighbouring countries — Togo, Upper Volta, Ivory Coast and even Nigeria. The immigrants came as cocoa-farm workers and for petty trading and diamond mining. But, when the economic fortunes of Ghana started to decline, Ivory Coast took its place as a major country of immigration in the area, and by 1975 had 1,426,000 immigrants (50 per cent of them from Upper Volta alone), or 21 per cent of its population.[1] Other important receiving countries like

---

1. More recent estimates indicate that the proportion of non-Ivorians in the Ivory Coast population is now close to 35 per cent. Europeans dominate high-level

Senegal and Sierra Leone had 300,000 and 67,174 immigrants in 1975 respectively. Elsewhere in the coastal states immigrants thrived, as the more recent influx into Nigeria emphasised.

The benefits and costs of emigration to the labour-exporters are well-known. The most widely recognised immediate benefit is the flow of remittances, which not only augments scarce foreign exchange earnings but also provides — especially in the poorer arid states — a potential source of additional savings and capital formation. Secondly, the migration of the unskilled and the unemployed, in addition to decreasing domestic unemployment, can even increase domestic employment to the extent that remittances stimulate demand for goods and services in the home country. Government welfare and social overhead expenditures may decrease, and the resultant savings can be channelled to investment (and hence produce more jobs). And thirdly, the forgone consumption of the migrated unemployed may also increase the savings of those hitherto supporting them.

To be set against these benefits to the labour-exporters are at least three sets of costs: first, the costs of loss of output (e.g. withdrawal of a few skilled workers at the margin can severely impair certain industries); secondly, the costs of additional education and training to replace the emigrating workers; and thirdly forgone output of the unskilled labour under training. As for the labour-importing countries, the gains of the exporters constitute their losses. Because of the danger of displacement of workers in the host-country, pressures on social services, particularly housing and medical and educational services, and sometimes an increase in crime, governments often intervene to bring the alien population to a culturally and economically tolerable level.

The purpose of this chapter is to show that Articles 2 (2*d*) and 27 (1,2) of the ECOWAS Treaty recognise the need to maintain, if not encourage, intra-regional migration as a way of rationalising and optimising resource use at the regional level. But the Treaty also realises its limits; hence the Protocol relating to Free Movement of

---

manpower occupations while non-Ivorian Africans occupy almost half the salaried jobs in the economy (see B.A. den Tuinder, *Ivory Coast: the Challenge of Success*, World Bank Report, Washington, 1978, p. 295). Obviously, this has its serious economic and political implications for a country now experiencing declining economic growth after its spectacular performance in the 1960s and the 1970s. Indeed, when in April 1981 some sixty unemployed Ghanaians were arrested and detained at Abidjan in a tiny room and all had died by the next morning, people saw in the callous handling of the affair a reflection of the natives' disgust with aliens.

Persons, Residence and Establishment clearly and meticulously stipulates how the delicate issue should be handled without excessive costs to either the labour-exporters or the importers.

2. The Treaty and Protocol relating to Free Movement of Persons, Residence and Establishment

As noted above, the Treaty of ECOWAS provides *by stages* (my emphasis) for the free movement of persons, residence and establishment. Article 2 (2*d*) of the Treaty says: 'The Community shall by stages ensure the abolition as between the Member-States of the obstacles to the free movement of persons, services and capital.'[2] Furthermore, Article 27 (1) confers the status of Community citizenship on the citizens of member-states, and enjoins member-states to abolish all obstacles to freedom of movement and residence within the Community. Realising the implications of this paragraph, Article 27 (2) categorically states: 'Member-States shall by agreements with each other exempt Community citizens from holding visitors' visas and residence permits and allow them to work and undertake commercial and industrial activities within their territories.'

From these two citations two things are clear: first, the Treaty provides for the free movement of persons, residence and establishment, and secondly, it recognises that these cannot be automatic — hence the qualifications implied in the use of the phrases 'by stages' and 'by agreements with each other'. The Treaty did not spell out in detail the stages to be followed or the time-table for the mutual agreements; these details were left to be filled in by a Protocol, which was later annexed to the Treaty.

The Protocol on Free Movement of Persons[3] was signed and annexed to the Treaty by the Authority of Heads of State and Governments at its annual meeting in Dakar on 29 May 1979, but it did not enter into force till a year later (20 May 1980) when it had been ratified by the necessary minimum of seven states (see pp. 147–51 and Table VI.5).

Although Article 2 (1) of the Protocol says that 'the Community citizens have the right to enter, reside and establish in the territory of

---

2. *Treaty of the Economic Community of West African States* (English version), Lagos, 1975, p.7.
3. The entire Protocol on free movement, residence and establishment is yet to be signed; only the limited free movement aspect is already in operation.

The treaty and protocol relating to free movement

member-states . . .', Article 2 (2) qualifies it as follows: 'The right of entry, residence and establishment. . .shall be progressively established in the course of a maximum transitional period of fifteen years from the definitive entry into force of this Protocol by abolishing all other obstacles to free movement of persons, residence and establishment.'

The Protocol left no doubt as to the process whereby the right of entry, residence and establishment was to be achieved. The operative time-table would be:
Phase 1 — Right of Entry and Abolition of Visa — 1980–5;
Phase 2 — Right of Residence — 1985–90;
Phase 3 — Right of Establishment — 1990–5.

The transition from one phase to another in the above schedule is not automatic.

It is planned that after the first phase has been effectively implemented for a maximum period of five years, the Trade, Customs, Immigrations, Monetary and Payments Commission, based on the experience gained from that implementation, should make proposals to the Council of Ministers for further liberalisation in the subsequent phases; also, that these phases would be dealt with in subsequent Annexes to this Protocol. Hence ECOWAS countries are still in the first phase of the General Principles on Movement of Persons, Residence and Establishment. And, *ipso facto*, any citizen of the Community who wishes to enter the territory of any other member-states is required under Article 3 (2) of the Protocol to possess a valid travel document and international health certificate. Such a person must enter through an official entry-point, and can stay for up to ninety days without a visa, but cannot stay longer without requesting and obtaining an extension. There is an escape clause (Article 4) empowering member-states to refuse admission into their territory to any Community citizen who comes within the category of inadmissable immigrants under its laws.

It is thus clear that most of the more recent population movements in West Africa took place outside the framework of the ECOWAS Treaty and Protocol. A study of the influx of Guineans into Sierra Leone and Liberia, of Voltaics, Malians and Ghanaians into Ivory Coast and of Ghanaians into Nigeria reveals some important facts. First, most of these immigrants entered their host-country — through the co-operation of immigration officials — without any valid papers at all and were therefore, *de facto* and *de jure*, illegal immigrants. Secondly, many of the immigrants entered their respective host-countries before the Protocol on free movement of persons was either signed or ratified. Thirdly, many of those who entered with or without valid papers exceeded the 90-days limit without obtaining permission for extension. And finally, a very high proportion of these immigrants entered the host-countries' territory through unofficial entry-points due to the vast and unpoliceable borders.

Those who had committed any or all of these offences could be classified as illegal immigrants, and Article 4 of the Protocol empowers a member-state to deal with such immigrants according to its laws. Thus member-states of ECOWAS have an inalienable right to act against immigrants who contravene not only the provisions of the Treaty and Protocol but also their own national immigration laws.

3. *Implications for net importers and net exporters of labour*

The spirit of the ECOWAS Protocol relating to Free Movement of Persons, Residence and Establishment is to foster the optimum utilisation of available resources, and even to foster development and faster economic growth at the regional level. In other words, the Protocol is consistent with the economic case for international migration.

To begin with, emigration in a labour-surplus economy would tend — all other things being equal — to increase the productivity and savings of those remaining behind while at the same time contributing to productivity in the receiving countries, especially where labour shortages have acted as a brake on economic expansion. In West Africa the development of plantations and forestry in the Ivory Coast has owed much to the Malian and Voltaic immigrant workers, and the same can be said of the growth of cocoa production in Ghana before and immediately after independence. Further afield, the emigration of many millions of people to the new territories of North America and the Antipodes from the 1840s to the 1870s enhanced the rapid economic expansion of these economies over the period.

Undoubtedly, immigrants can make positive contributions in other ways. The presence of immigrants from ECOWAS countries in Nigeria has shown that is it not altogether an unmitigated evil. They could provide a source of cheap industrial labour, especially in construction, hotel service and other manual occupations. Indeed, some Nigerian state governments, especially Lagos, rely heavily on an estimated 35,000 Ghanaian teachers to carry out their educational policy. Professional groups like doctors, nurses, lawyers, accountants, engineers and pilots supply additional critical skills while the technical groups supply needed middle-level manpower.

But even allowing for these positive contributions, one important question needs to be resolved first: what is the absorptive capacity of the labour-receiving economy? Can it cope with an ever-increasing number of foreign immigrants? It was in recognition of the limited capacity of any economy to absorb new entrants that the Treaty and Protocol prescribed specific conditions to ensure that the entry of immigrants should be regulated to the mutual advantage of both the importer and exporter. Unfortunately, the inability of the immigrants to adhere to the provisions of the Treaty and Protocol has created problems for all the parties concerned.

First, there is a politico-economic limit to the size of immigrant population which any society can tolerate, this being a function of the country's size, the absorptive capacity of its economy, and its socio-cultural and allied factors. Once it is exceeded, tensions begin to build up and if not checked can give rise to tragic developments such as were seen early in 1983 in India's north-eastern state of Assam.[4]

Secondly, an increase in labour supply — whether from domestic sources, through immigration, or both — presupposes a corresponding increase in the growth of job opportunities. Otherwise the additional labour — mostly young people — flowing into the job market will remain unemployed, with obvious adverse socio-political consequences. Since the 1970s the economies of most West African countries have been declining in real terms and, with the concurrent world-wide recession, unemployment has grown. The capacity of their economies to cope with large immigrant populations was overtaxed, hence the partial expulsion of Guineans from Sierra Leone in December 1982, the expulsion of illegal unskilled West African immigrants from Nigeria in February 1983, and the uneasy truce between natives and immigrants in the Ivory Coast. It is clear that for as long as the presence of foreign immigrants seriously threatens the job security of the natives, the weapon of expulsion will always be held in reserve. Among other things, it is a convenient means for governments to improve their popular image at home. But the cost of expelling immigrants in terms of international image, material support and massive foreign exchange transfers, even if a necessary evil, can be considerable.

Thirdly, the presence of a large immigrant population has other costs. It can be the cause of regular large outflows of foreign exchange through remittances, especially in countries like Nigeria where foreigners are allowed to transfer up to 50 per cent of their

---

4. See *The Economist* (London), 29 Jan. 1983, and the BBC 'World News Report, 24 Feb. 1983.

gross earnings. Similarly, the demand for social and welfare services may entail some additional investment by the Government. Housing problems deserve special mention here, if only because immigrants tend to concentrate in the urban centres which are already hopelessly overcrowded.[5] So they not only influence the rate of urbanisation in their host-countries but exacerbate the problem of overcrowding. This comes out clearly in Table VI.1. In the 1960s the major host-countries for immigration in West Africa — Ghana, Ivory Coast, Liberia and Senegal — showed relatively high rates of urbanisation. And the trend continued in the 1970s with few exceptions. This same group of countries also exhibited a high rate of labour supply with its attendant problems.

Finally, some unemployed foreign immigrants show little inhibitions in joining criminal gangs in the host-country where, being unknown to the natives, they feel safe. Nigeria provided a paradise for this group of immigrants since it already had the largest number of armed robbers in West Africa. Little wonder therefore that investigations revealed that certain immigrants in Nigeria were active participants in the religious riots of 1980 and 1982 and were among the guilty in known cases of armed robbery. Prostitution reached its peak at the height of foreign migration to Nigeria in recent times. There could also be political costs, as and when politicians recruit and train immigrants for motives considered to be against civilised and national interests. Less specifically, there is the social and cultural cost of the exposure of impressionable natives to ill-motivated immigrants.

4. Policy Issues and Recommendations

We can see in all the above a number of policy issues. First and foremost, Nigeria's expulsion of illegal immigrants is not in conflict with the ECOWAS Protocol on free movement of persons. In fact, the Executive Secretary of ECOWAS defended the action of Nigeria in the ECOWAS context.[6] The influence of the 90-day-visit-without-visa privilege for ECOWAS citizens on the influx of immigrants cannot have been as great as some have maintained; Chadians and Cameroonians, who are non-citizens of the ECOWAS community, were also present in large numbers. The key lies with the 'porous' borders of African countries. Adequate border policing is, more than anything else, the answer to checking international migration in the sub-region.

5. A recent survey conducted by the Nigerian Institute of Social and Economic Research reveals that the urbanisation rate of aliens is three times higher than that of the natives (see NISER Surveys, 1982).
6. See *West Africa*, 7 Feb. 1983.

Another issue turns on the skills which the immigrants possess and their usefulness to the host-countries. Most immigrants are unskilled and semi-skilled workers whose services are not considered to be of critical importance, since those services are probably in abundant supply in the receiving country. This means that unregulated acceptance of this group would entail a degree of displacement of workers in the host-country, any level of which is bound to provoke the hostility of the natives, and lead ultimately to punitive measures against the aliens. To avoid this, labour-importing countries should accept immigrants only for sectors and subsectors of the economy where the native workforce is unlikely to be displaced. There is also the attitude of the labour-exporting countries to the problem of mass exodus of their citizens. Migrants always want to leave a weak or collapsing economy for the more promising ones, and therefore in West Africa, where most of the economies are weak, the few stronger ones like Nigeria, Ivory Coast, Senegal and, to a small extent, Liberia and Sierra Leone will always be under pressure. It is in recognition of this that the ECOWAS Protocol lays down a strict procedure for free movement of persons, and it expects the member-countries to regulate the rate of exodus of their nationals. This is important in the interest of all.

Furthermore, when it becomes irresistible to expel some of the immigrants, the exporting countries involved should be informed in advance through the diplomatic channels. As a prelude to the expulsion order, adequate preparations must be made by all concerned to ensure an orderly exit with the minimum of human suffering. Also, the exit deadline must be reasonable. In early 1983, Nigeria gave an alien population estimated at about 2 million[7] two weeks to leave the country. Although this was later extended by one month for the skilled workers, the two-weeks deadline was inadequate and the handling was haphazard. It is of course true that Ghana's Aliens Compliance Order of 1969 gave all immigrants in their midst fourteen days to leave the country, but two wrongs do not make a right. It seems that the anti-Nigerian reaction on the subject in the Western press was based largely on the timing and management of the affair. But this went to the point of questioning the legitimacy of the action itself, despite the conclusive contrary evidence of the ECOWAS Protocol.

---

7. The estimates range from a low of 1 million to a high of 2 million, but of course nobody knows the exact number. At best the estimates are speculative but the number involved was undoubtedly large, and most were Ghanaians. It has, indeed, been reported that by the middle of February some 1.3 million Ghanaians had returned to Ghana from Nigeria (see *The Economist*, 19 Feb. 1983).

The aims of ECOWAS cannot be achieved without enforcement of the principle of free movement of peoples within the sub-region. But, given the dissimilarities in the economies of the sub-region, this must be regulated, which means that the provisions of the Protocol on free movement of persons must be strictly adhered to. Anything different would be counter-productive. Thus, in order to avoid occasional expulsions and their unpleasant consequences, the following policy is recommended.

(*a*) ECOWAS member-countries should always ensure that their citizens adhere strictly to the provisions of the Treaty and the protocols attached to it. To effect this, it might be advisable in the future for West African governments to issue exit visas to their citizens wishing to migrate to other parts of West Africa. This, clearly would not only ensure adherence to the Treaty and Protocol on free movement but also help to restrain any mass exodus.

(*b*) West African countries should strengthen the policing of their borders. This would necessarily entail measures to deter the border guards and immigration officials from collaborating with illegal immigrants and the reorganisation of the existing inefficient immigration departments in several countries.

(*c*) Labour-importing countries should try to accomodate their immigrant labour force in such a way as to avoid displacement or clashes with the local people. A system of periodic or annual registration of aliens (as is done routinely in places like United States and Britain) might be very useful here. Through this means the geographic (i.e. urban overcrowding) and occupational concentrations of aliens could be handled.

Lastly, where and when expulsion becomes inevitable, the countries whose citizens are affected should not only be informed in advance, but there should be joint preparation and execution of the expulsion order, if possible within the framework of ECOWAS. To act otherwise might provoke retaliatory measures, and thus threaten the very existence of the Community itself.

SELECTED REFERENCES

Books

Amin, S., *Neo-Colonialism in West Africa*, Harmondsworth: Penguin Books, 1973.

Balassa, B., *The Theory of Economic Integration*, London: George Allen and Unwin, 1962.

——, et al., *The Structure of Protection in Developing Countries*, Baltimore: Johns Hopkins Press, 1971.

Barzanti, S., *The Underdeveloped Areas within the Common Market*, Princeton University Press, 1965.

Brandt, Willy (Chairman), *North-South: a Programme for Survival*, London: Pan Books, 1980.

Brown, A.J., 'Should African Countries Form Economic Unions', in Jackson (ed.), *Economic Development in Africa*, Oxford: Basil Blackwell, 1965.

Carnoy, M. *Industralization in A Latin American Common Market*, Washington DC: Brookings Institution, 1972.

C.E.D., Blough, R. and others, *Regional Integration and the Trade of Latin America*, Committee for Economic Development, January 1968.

Church, R.J.H., *Environment and Policies in West Africa*, Princeton: Van Nostrand, 1963.

——, *Some Geographical Aspects of West African Development*, L.S.E., January 31st, 1966.

Demas, W.G., *The Economics of Development in Small Countries with Special Reference to the Caribbean*, Montreal: McGill University Press, 1965.

Diejomaoh, V.P., and Associate (eds), *Industrialization in the Economic Community of West African States (ECOWAS)*, Ibadan: Heinemann Educational Books (Nigeria), 1980.

Dike, K.O., *Trade and Politics in the Niger Delta: 1830–1885*, Oxford: Clarendon Press, 1956.

Dosser, D., et al., *A Theory of Economic Integration for Developing Countries*, London: George Allen and Unwin, London, 1971.

Ezenwe, Uka I., 'Economic Integration in West Africa', unpublished Ph. D. thesis, University of St Andrews, Scotland, 1976.

Gardiner, R.K., 'African Economic Development' (mimeo), lectures delivered at the University of Stratchlyde, June 1971.

Green, R.H. and K.G. Krishna, *Economic Co-operation in Africa, Retrospect and Prospect*, Nairobi: Oxford University Press, 1967.

Green, R.H. and A. Seidman, *Unity or Poverty? The Economics of Pan-Africanism*, Harmondsworth: Penguin Books, 1968.

Hargreaves, J.D., *Prelude to the Partition of West Africa*, London, 1963.

Hazlewood, A. (ed.), *African Integration and Disintegration*, London: Oxford University Press, 1967.

——, A., *Economic Integration: the East African Experience*, London: Heinemann, 1975.

Hilton, R. (ed.), *The Movement Toward Latin American Unity*, New York, Praeger, 1969.
Hirschman, A.O., *The Strategy of Economic Development*, New Haven: Yale University Press, 1958.
Hunter, G., *The Best of both Worlds? — a Challenge on Development Policies in Africa*, Oxford University Press, 1967.
IADB, *Multinational Investment in the Economic Development and Integration of Latin America*, Bogota; IADB, April 1968.
IMF, *Surveys of African Economies*, vol. 3, Washington DC, 1970.
Kahnets, F. et al., *Economic Integration among Developing Countries*, Paris: OECD, 1969.
Kobe, S., *Transport Problems in West Africa*, Paris: OECD, 1967.
Little, I.M.D. and J.A. Mirrlees, *Manual of Industrial Project Analysis in Developing Countries*, vol. II, Paris: OECD, 1968.
Machlup, F. (ed.), *Economic Integration: World-wide, Regional, Sectoral*, London: Macmillan, 1976.
Mutharika, B.W.T., *Toward Multinational Economic Co-operation in Africa*, New York: Praeger, 1972.
Okigbo, P.C., *Africa and the Common Market*, London: Longmans, 1967.
Organisation of African Unity, *Lagos Plan of Action for the Economic Development of Africa: 1980–2000*, Addis Ababa, 1981.
Pearson, Lester (Chairman), *Partners in Development*, London: Pall Mall Press, 1969.
Peterec, R.J., *Dakar and West African Economic Development*, New York: Columbia University Press, 1967.
Plessz, N.G., *Problems and Prospects of Economic Integration in West Africa*, Montreal: McGill University Press, 1968.
Renninger, J.P., *Multinational Co-operation for Development in West Africa*, Oxford: Pergamon Press, 1979.
Robson, P., *Economic Integration in Africa*, London: George Allen and Unwin, 1968.
——, *Current Problems of Integration*, UNCTAD, 1971.
—— (ed.), *International Economic Integration*, Harmondsworth: Penguin Books, 1971.
Scitovsky, T. et al., *Industry and Trade in Some Development Countries*, a Comparative Study, London: Oxford University Press (for OECD), 1970.
Thompson, V., *West Africa's Council of the Entente*, Ithaca: Cornell University Press, 1972.
Wionczek, M.A. (ed.), *Latin American Economic Integration*, New York: Praeger, 1966.
United Nations, *Report on the Alternatives of Association between the Gambia and Senegal*, New York, March 1964.
UNCTAD, *Payments Arrangements among Developing Countries for Trade Expansion*, TD/B/C/24, July 1966.
The Treaty of the Economic Community of West African States, Lagos, May 1975.
World Bank, *Accelerated Development in Sub-Saharan Africa: An Agenda*

for Action, Washington DC, August 1981.
——, *World Development Report*, 1981, Washington DC, 1981.

Articles

Adedeji, A., 'Prospects of regional economic co-operation in West Africa', *Journal of Modern African Studies*, 1966.
Allen, R.L., 'Integration in less-developed Areas', *Kyklos*, XIX, 1961.
Balassa, B., 'Towards a Theory of Economic Integration', *Kyklos*, 14 (1) 1961. (??)
Bretton, H.L., Report of a seminar on Economic Co-operation in Africa, held at Nairobi, *Journal of Modern African Studies*, May 1966.
Carlson, S., 'Some Notes on the Dynamics of Economic Integration', *Swedish Journal of Economics*, 72, 1, January 1970.
Cox-George, N.A., 'Economic Structures of West African Countries', *The Nigerian Journal of Economic and Social Studies*, 5, 1, March 1963.
Dosser, D., 'Customs Unions, Tax Unions, Development Union', Institute of Social and Economic Research and Department of Economics, University of York, Reprint Series, 145.
Ezenwe, Uka I., 'The Rationale of Economic Integration in West Africa', *Intereconomics*, Hamburg, April 1975.
——, 'The Prospects of Improving Trade and Development in the Third World in the 1980s' in A.M. Osoba (ed.), *Developing Countries and the New International Economic Order*, Ibadan: NISER, 1979.
Hazlewood, A., 'The East African Common Market: Importance and Effects', *Bulletin of the Oxford Institute of Economics and Statistics*, February 1966.
Jaber, T.A., 'The Relevance of Traditional Integration Theory to LDCs' (Review article), *Journal of Common Market Studies*, IX, March 1971.
Kravis, I.B., 'Trade as a Handmaiden of Growth: Similarities between the 19th and 20th Centuries', *Economic Journal*, December 1970.
——, 'Economic Grouping in Africa' in C. Legum and J. Drysdale (eds), *Africa Contemporary Record*, 1969-70, 1970.
Mead, D.C., 'The Distribution of Gains in Customs Unions between Developing Countries', *Kyklos*, 21 (4), 1968.
Meier, G.M., 'Effects of A Customs Union on Economic Development', *Social and Economic Studies*, 9, 1-4, 1960.
Onitiri, H.M.A., 'Towards a West African Economic Community', *N.J.E.S.S.*, 5, 1, March 1968.
Sakamoto, J., 'Industrial Development and Integration of Underdeveloped Countries', *Journal of Common Market Studies*, 7, 1969.
Scitovsky, T., 'International Trade and Economic Integration as a means of over-coming the Disadvantages of a Small Nation' in E. Robinson (ed.), *The Economic Consequences of the size of Nations*, London: Macmillan, 1963.

U.N. documents

Strategy for Development in Africa in the 1970s, E/CN.14/493/Rev.1.
Fiscal Compensation and Economic Integration, TD/B/322, 1970.
(UNCTAD), *State Trading and Regional Economic Integration*, TD/B/436, 1973.
(UNCTAD), *The Distribution of Benefits and Costs in Integration among Developing Countries*, TD/B/394, 1973.
ECA, *Some Institutional Aspects of African Economic Co-operation*.
——, *Economic Co-operation in Africa*, E/CN.14/UNCTAD/11/4.
——, *Report of the West African Sub-regional Conference on Economic Co-operation*, E/CN.14/399, 1967.
——, *Economic Co-operation and Integration in Africa: Three Case Studies*, ST/ECA/109, 1969.
UNCTAD, *Economic Co-operation and Integration among Developing Countries*, 1, 19 May 1976, TD/B/609.
——, *Economic Co-operation and Integration among Developing Countries*, 2, 20 May 1976, TD/B/609.

INDEX

Abidjan, 78, 135, 191; Treaty of (1973), 81, 82-3
Accra, 123, 143
Acheampong (Ghana head of state), 135
African-Caribbean-Pacific Group of States (ACP), 34
African Development Bank, 68, 71, 156
African Development Fund, 68, 71
Afrique Occidentale Française, *see* French West African Federation
agriculture, agricultural production: 18, 20, 82, 97, 103, 138, 169, 176; primary, 17, 138; subsistence, 19; crop finance, 89; exports, 134
agro-industrial enterprises, 127, 139-40
aid, foreign: 155-6, 181, 184-5; per Lomé Convention, 34-5
air traffic, aviation, airways: 140, 157, 169, 174
Akosombo dam (Ghana), 134
Algeria, 24, 37
America, North, as market for African goods, 26, 49, 138
Amîn, Idi, 9
Andean Common Market, 61n
Anglophone countries: 38, 70, 83, 126, 153; cultural relations with Francophones, 41, 84-5, 126, 153; attitude to economic co-operation, 66n
animals, *see* livestock
animal products, 23-4
Arab Common Market, and Egypt, 32n, 60, 151
Argentina, 53

'backwash' effects, 60
balance of payments, 55, 59
Bamako: 108, 145; Agreement (1970), 80, 81, 84
bananas, 181
Banque Centrale des Etats de l'Afrique de l'Ouest (BCEAO), 38, 86-90, 92, 104, 123: currency issue, 88; as deposit for reserves, 88; credit creation, policy, 88-90
Banjul, 112, 117, 135
bauxite, 17, 110, 135

beef, 139, 181
Belgium, 132
Benin (Dahomey), 1, 3, 38, 39, 84, 138, 143, 170, 178n, 188; and UDEAO, 72, 74; and CEAO, 81, 83; and Algeria, 101; and Togo, 88; and BCEAO, 89, 91; and Entente, 93, 94, 96, 97, 99n; and R. Niger Commission, 120, 121; and WACH, 123; exports, 174; emigration, 180; socialist system, 144
Benin, Bight of, 36
Berlin: Congress, 2; General Act of (1885), 120
beer, production, 20, 98
beverage industry, 20, 21
Bhambri, R.S., 47
bicycle assembly, 23
Bissau, 135
Bolivia, 53, 61n
Brandt Commission Report, 185, 186
Brazil, 53
breweries, 23
Britain: 2, 4, 38, 84; aid to Gambia, 111n, 112n, 119; and aliens, 198
British West Africa, 5, 67, 190

Cameroon: and R. Niger Commission, 120; and Chad Basin Commission, 121; road links, 135; oil production, 182; emigrants, 196
Canada (aid), 121
Canadian International Development Agency (CIDA), 100
Cape Verde Islands, 131, 172, 178n
capital goods industries, 21, 186
capital supply, 32
Caribbean Free Trade Association, 151
Casamance (Sen.), 111, 112
cassava, 138
Castillo, C.M., 152
cattle, 138
Cayar (Sen.), refinery, 134
cement industry, 20, 98, 101n, 106, 134
Central African Customs and Economic Union (UDEAC), 58, 61, 151
Central American Common Market, 52, 60, 79n, 151
Central Bank of the West African States,

203

Index

see Banque Centrale
Central Banks, Committee of West African, 141
CFA franc, 4–5n, 31, 38, 86
Chambers, Sir Paul, 23
Chad (Lake) Basin Commission (CBC), 68, 70, 121–2
Chad: and R. Niger Commission, 120; road links, 135, emigrants, 196
chemical industry, 20, 104
Chenery, H. and M. Syrquin, 20
Church, R.J.H., 138
civil service, 173
class conflict, 135
classification of organisations, 70–1
clothing, 20
coal, 17
coastal highway, W. African, 161
cocoa, 17, 26, 134, 181, 190, 194
Cocoa Producers' Alliance, 68, 70, 71
coffee, 181
Collin, Jean, 112
Common Market theory, 11, 46, 52
compensation: strategy, 41; fund to adjust costs/benefits in ECOWAS, 168
Conakry, 108
consumer goods industries, 9, 21, 28, 47, 186
copper, 106, 110, 181
Costa Rica, 54, 60, 79n
Cotonou, 97
cotton, 181
Council for Mutual Economic Assistance (CMEA), 42
Court of Arbitration of French Community, 74
crime (and immigration), 191, 196
cultural affairs, 127, 128
currency: over-valuation, 51; union, 67n, 141–2, 168
customs duties: revenue effects, 58–9, 62, 79; elimination, 127
customs union: 3, 5, 11, 43–5, 50, 122; benefits from, 7ff.; theory of, 42–4, 52, 54, 55, 72, 73, 75; in ECOWAS, 133
Customs Union of West African States (UDEAO): 23, 31, 58, 68, 71, 72, 75–80, 81, 86, 92; impact on trade, 77–9; Convention of 1959, 53, 73–4; Convention of 1966, 75–6; collapse, 53, 79–81

Daddah, Pres. Ould, of Mauritania, 108, 126
Dakar: 78, 79, 102, 109, 116, 117, 135, 145, 162; Agreement of May 1979, 169
David, J.E., 80–1, 82
defence: 99n, 116, 172; Senegambia, 116–17, 119
diamonds, 17, 123, 134, 190
Diori, Pres. Hamani, of Niger, 84, 93
disparities in development levels, 78
diversion, trade, 45–7, 77n
Dosser, D., 55
drought, 136, 173, 176
dual nationality, and integration schemes, 95, 133
duty, countervailing, 182
dynamic effects of integration, 45

East Africa, 140
East African Economic Community (or Common Market), 9, 51, 53, 58, 60–1, 78–9n, 82, 151, 155
Economic Commission for Africa, see under UN
Economic Community for Cattle and Meat (Entente), 98–9
Economic Community of West African States (ECOWAS): 14, 31, 33, 35, 37, 39, 41, 46, 60–4, 70, 80, 101, 120, 125, 126–55 passim, 161–2, 167, 169, 183–5, 186–7, 188–9, 191, 193–4, 196–8; foundation, 126; aims, 126–7; organisational structure, 127–31; and import duties, 59n; common payments system, 141; exchange agreement, 141–3; regional company law, 162; (non-)ratification of Protocols, 149, 151; members' unwillingness to fulfil obligations, 147
ECOWAS Secretariat: 141, 196; relations with other organs in grouping, 167
ECOWAS Treaty: 33, 37, 38, 59n, 62, 64, 83, 85, 124, 130–2, 141, 151, 153, 167, 191; and tariff concessions, 60–1; and customs union, 133; and free mobility of labour, 133, 192
economic union, integration: 11; necessary for industrialisation, 21; increase in members' buying strength, 23; reduces vulnerability of individual states, 30; general benefits, 50, 62–3, 169

Index

Edel, M., 60
education, 4, 101, 107, 118, 157, 173, 183, 191, 194
Egypt, 32n, 52, 60
El Salvador, 54, 60, 79n
electricity resources, 17
engineering industry, 140
Entente states, 9, 33, 49, 53, 68, 70, 71, 81, 92–100, 101; Council, 92–3; Mutual Aid and Guarantee Fund, 98; Solidarity Fund, 93–4
enterprises, 'Community' (ECOWAS): conditions for approval, 163–4; investment guarantees, 164–7
Equatorial Customs Union, 75
Europe, Western, as market for African goods, 26, 49, 138
European Economic Community (EEC), and ACP states, 33–5, 50, 71, 82, 84
exchange, foreign: control, 51; agreement (ECOWAS), 142; shortages, 174; remittances by emigrants, 180, 191, 195
exports: cash crop, 24, 25; primary, 26, 152, 175, 178, 184, 186; manufactured, 178; common policy, 82; to industrial countries, 138, 183; level in region, 174–5, 176, 178; promotion, 174, 181
extractive industries, 32
Eyadema, Pres., of Togo, 83, 126

fiscal policy: (transfers) 8, 53, 54, 101, 114
fishing, 18, 82, 122, 139
flour mills, 104
food: industry, trade, 26, 28, 103, 176; imports, 178–9; technologies, 140 and n; reserves, 169; supply and demand, 181
Food and Agriculture Organisation of the UN (FAO), 113, 115, 122, 157
forestry, 18, 194
Fort Foureau, 122
Fotokol, 122
forward linkage of W. African manufacturing, 21
Fourah Bay, 4
France: 2, 3, 4, 35, 83, 85; Treasury, 86, 88; and Entente, 99n; support for CEAO, 144, 153
Francophone West African countries: 4, 33, 81–2, 83–5, 126, 153; and economic co-operation, 67; special relations with France, 35; *see also* West African Economic Community
free trade, 32, 44
Freetown, 123, 162
French Equatorial Africa, 7
French West African Federation, 4, 6, 7, 72, 190
Fund for Co-operation, Compensation and Development (FCCD), 64

Gambia, The, 3, 38, 39, 131, 178n, 188; co-operation with Senegal, 111–20; and Mano River Union, 123; and WACH, 123
Gambia River and basin, 112, 114–15, 118
Gaya (Niger), 121
General Agreement on Tariffs and Trade (GATT), 50, 182
Ghana: 3, 26, 28, 38, 39, 60, 83, 84, 138, 143, 178; manufacturing, 20, 23, 32; heavy industry, 46; road and rail links, 36; growth rate, 51; import duties, 59n; and Entente, 94, 96, 99, 101; and Togo, 101; and WACH, 123; and ECOWAS, 131, 134; indigenisation policies, 153; exports, 174; oil production, 182; immigration, 190, 194, 196, 197; Aliens Compliance Order, 197; emigration, 135–6, 180, 191n, 193, 194, 197n
Gross Domestic Product (GDP), 18, 20, 32, 171, 178, 180
Gross National Product (GNP), 45–6, 49, 60
gold, 1, 17, 134; as security, 89
government as employment sector, 22
Gowon, Gen. Y., 83, 126
Green, R.H., 8
groundnuts, 19, 26, 112, 122, 138, 181; oil, 181; Senegal, 114
Guatemala, 54, 60, 79n
Guinea, 7, 33, 39, 72, 100, 144, 161, 172, 178n, 188; and OERS, 101, 104, 105, 108–9; currency, 104; and ECOWAS, 143; GDP, 171; and R. Niger Commission, 120
Guinea-Bissau, 108, 131, 178n
guinea corn, 138

Haas, E.B., 137

Haberler-Marshall, 44
harmonisation: of economic policy, 126; industrial, 168; agricultural, 169; within Entente, 95, 98; within ECOWAS, 134
Hazlewood, A., 6
health: public, 97; regulations, 182–3
Hodgkin, T., 7
Honduras, 54, 60, 79n
Horton, Dr Romeo, 167
Houphouet-Boigny, Pres. F., of Ivory Coast, 84, 85, 92–3, 126, 133; Convention (1965), 94–5
hydro-electric works, Gambia, 118

Ibadan University, 4
import substitution, 20n, 28, 46, 59, 62, 186
imports: 174–5; restrictions 51
income effect, 47
India, 180, 195
Industries Cotonnières du Dahomey (ICODA), 97
industry, industrialisation: 21, 24, 32, 33, 54, 56, 82, 104, 186, 187–8; allocation of regional --, 63–4; inter-union industrial location 161; contribution to GDP, 138; harmonisation of development, 77–8; and OERS, 104, 106; inter-regional co-operation, 106, 127; export-oriented, 132; independent national, in Nigeria, 151
Industry, Agriculture and Natural Resources Commission (IANRC), 64
inflation, 181
infrastructure: 31; social, 97
institutions: post-colonial, adaptation, 173; regional (ECOWAS), 183–4
integration, see List of Contents (pp. vii–ix)
Inter-African Coffee Council (IACO), 68, 70, 71
International Civil Aviation Organisation (ICAO), 36
investment policy, 62
iron, 26, 110, 161, 181, 188
irrigation, 115, 118, 122, 138
Ivory Coast: 5, 8–9, 28, 38, 60, 84, 133, 134, 138, 144, 172, 178; and Senegal, 9, 78, 92; manufacturing, 20, 23, 32, 33; road and rail links, 36–7; and UDEAO, 53, 72, 74; and CEAO, 80, 81; and Entente, 53, 92–5, 96, 99n, 100, 101; arrangements with Upper Volta, 77; and BCEAO, 89; rise in assets, 91; and R. Niger Commission, 120; and ECOWAS, 131; and Lebanese, 170; imports, 174; agricultural production, 176; immigration, 180, 190–1 and n; oil production, 182; and Guinea, 188; and Ghana, 190

Jawara, Pres. Sir Dadwa, of Gambia, 128
Johnson and Cooper, 54

Kandiji (Niger) dam, 134
Kaya, Paul, 95
Keita, Pres. Modibo, of Mali, 107
Kenya, 24, 37, 53, 78–9, 155
Kitamura, H., 47
Kuznets, S., 26

labour: supply, 22, 32, 60, 132, 180, 194–5; productivity, 46–7; intra-ECOWAS mobility, 63–4, 133, 134–5, 191
Lagos: 85, 123, 126, 135, 194; Plan of Action, 186–7
Lake Chad Sprinkler Irrigation Scheme, 122
landlocked states, 28, 35, 79, 85, 94, 115, 134
Latin American Free Trade Area (LAFTA), 32n, 51, 53, 58, 151
lead, 181
Lebanese communities in W. Africa, 170
leather industry, 106, 138
Legon, University of, 4
Liberia: 14, 26, 28, 60, 131, 136, 151, 161, 167, 178, 188; agriculture, 18–19; and Mano River Union, 122–3; and WACH, 123; foreign trade, 175; immigration, 180, 193, 196
licensing, industrial, advocated, 65, 162, 168
Linder, S., 52
linkages of W. African manufacturing, 21, 54
livestock production: 28, 82, 169, 173; and OERS, 106–7; and OMVS, 110
Little, Scitovsky and Scott, 30
Lomé, 97
Lomé Convention: 31, 33, 34, 135, 68, 71; — II, 34

Index

Maghreb countries, 84
Maiduguri, 122
maize, 110, 122, 138, 139, 181
Makower, H., and G. Morton, 48
Malanville (Benin), 121
Mali: 39, 104, 136, 138, 178n; road and rail links, 36, 135; and tariffs, 49; and UDEAO, 72-3; and CEAO, 80; and UMOA, 86; and OERS, 101, 104, 105, 107; and USAID, 106; and OMVS, 109-10; and R. Niger Commission, 120; and WACH, 123; and ECOWAS, 134; GDP, 171; exports, 174; emigration, 180, 193, 194
Mali Federation, 92
Malick, Mr (Upper Volta), 100
Manantali (Mali), proposed dam, 103-4, 134
Mano River Union, 68, 70, 122-3; and ECOWAS, 144, 145-6
marble quarrying, 97
Mauritania: 14, 26, 38, 39, 91, 104, 111n, 131, 136, 138; and UDEAO, 72-3, 74, 75; and CEAO, 80, 84; and Senegal, 88; and OERS, 101, 105, 106, 108; and ECOWAS, 143; foreign trade, 174; and OMVS, 109-10
meat production, 97, 100, 110, 138, 139
metal processing, 98, 104, 106
meteorology, 136, 157
Mexico, 53
migration, 4, 16, 128, 169, 172, 180, 190-8; illegal, 193-4, 195, 198
milk production, 110
millet, 138
mining, 138, 161
monetary policy, 54, 85ff, 101, 104, 114, 127
monetary system, W. Africa: 3, 4, 11; Francophone, 30; *see also* West African Monetary Union
monopoly, industrial, 63
Monrovia, 122
multinational firms, 161, 167
Multinational Trade Negotiations, 183
Mutual Aid and Guarantee Fund (Entente), 95, 98

N'Djamena, 121, 135
Niamey: 85; Act of, 120-1
Nicaragua, 54, 79n
Niger: 14, 38, 39, 49, 84, 88, 91, 136, 138-9, 178n, 188; manufacturing, 20; roads, 36-7, 135; and UDEAO, 72-3; and CEAO, 80, 81; and Entente, 92-3, 94, 96, 97, 99n; and Nigeria, 101, 143; and R. Niger Commission, 120, 121; and WACH, 123; and ECOWAS, 134; GDP, 171; exports, 174; emigration, 180
Niger River, 157; Commission, *see* River Niger Commission
Nigeria: 38, 39, 60, 83, 143, 153, 16, 178, 188; population, 14; road and rail links, 36, 101n, 135; Northern, 16; oil industry, 17, 26, 28, 59, 171; agriculture, 18-19; heavy industry, 46, 138, 151; manufacturing, 20, 32; bicycle assembly, 23; and neighbouring countries, 101; and tariffs, 49, 72; imports, 174; import duties, 59n; and R. Niger Commission, 120; and WACH, 123; and ECOWAS, 131; drought in North, 136; indigenisation policy, 153; GDP, 171; immigration, 180, 191, 193, 194, 195-6, 197
nitrates, 104
Nkrumah, Kwame, 6, 107, 136, 137n
Nouadhibou (Maur.), refinery, 134
Nouakchott, 108, 109, 135
Nyerere, Pres. J., of Tanzania, 9

'off-setting' (compensatory) tax, 61-2, 64
oil: industry and products, 17, 98, 181; Nigeria, 17, 26, 28, 151; exporters, 177-8; importers, 174, 178, 180, 181-2, 184, 187
Organisation for the Development of the Senegal River (OMVS), 109-10, 144
Organisation of African Unity (OAU), 35, 67, 68, 70, 71, 100
Organisation of Co-ordination and Co-operation for the Fight against the Major Endemic Diseases (OCCMED), 68, 71
Organisation Commune Africaine et Malagache (OCAM), 68, 70, 71
Organisation of Petroleum Exporting Countries (OPEC), 187
Organisation of Senegal River States (OERS), 68, 70, 81, 101-4, 106-10
Ouattara, Dr A.D., 167
Ougadougou: 97, 98; headquarters of CEAO, 82

palm kernels, 17
palm oil, 17
paper manufacture, 104
Paris: UDAO Conference, 75; headquarters of BCEAO, 86
payments: problems, 31; advantages of union, 37–40, 141, 168
Pearson, Lester, 185
Pennsylvania, Univ. of, report, 185
petrochemicals, 106
petroleum, *see* oil
pharmaceutical products, 106
phosphates, 110, 181
plasterworks, 106
polymers, 106
Pompidou, Pres., of France, 83
population: distribution and growth, 14–16, 176; free movement of, 127, 133, 149, 169–70, 188, 190; *see also* Protocol on Free Movement . . .
Portugal, 108 and n
postal work, 157
poultry, 106–7
prices: 181; integrated, 168–9
productivity levels, 46–7, 186, 194; *see also* labour
protectionism, trade, 33, 35, 50, 57, 182–3, 186
Protocol on Free Movement of Persons, 192–3
public sector, deficiencies, 173–4, 177

quantitative restrictions, 183

rail links, 49, 169
reafforestation, 122
rice, 110, 138, 176
river basins, development, 136, 169
River Niger Commission (RNC), 68, 70, 71, 120, 121
Robson, P., 24, 26n, 115
root crops, 138, 139
rubber, 181

Sahel, 99, 136; international research on, 169; drought, 173
St Germain-en-Laye Convention (1919), 120
Sanyang, K.S., 119
savings: 191, 194; domestic, 186
Senegal: 3, 5, 26, 28, 32, 38, 39, 60, 78, 84, 92, 104, 138, 151, 172, 178; agriculture (groundnut), 18–19; and UDEAO, 53, 72–3, 75; and Guinea, 188; and Upper Volta, 77; and CEAO, 80, 81; and Mauritania, 88; decline in assets, 91; and OERS, 101, 105, 106, 108, 109; and the Gambia, 111–20, 134; and WACH, 123; foreign trade, 175; immigration, 180, 191, 196
Senegal River basin, 103–4, 157
Senegambian integration: 68, 70, 110–20; treaty of 1967, instruments, 117; defence and foreign affairs, 116–17; cultural and sports co-operation, 118; economic benefits, 114–16
Senengue (Mali) dam, 134
Senghor, Pres. L., of Senegal, 84n, 85, 108, 153
Sierra Leone: 14, 28, 38, 39, 60, 134, 136; and Mano River Union, 122–3; and WACH, 123; and ECOWAS, 135; imports and exports, 174; immigration, 180, 191, 193, 195
slave trade, 2, 16
smuggling, 49, 55, 112, 114, 116, 119, 123, 134–5
social services, and mass immigration, 191, 196
Société Dahoméenne du Kenaf (SODAK), 97
Solidarity Fund (Entente), 93–4
South Africa, 100
South America, 53–4
South-east Asia, 50
specialisation in production, intra-zonal, 46, 131, 134, 169
sport, 118
statistics, compilation of, 82
subsistence sector, 32
sugar: 110, 122, 138, 181; refining, 104
Switzerland, 131
Syria, 52

Tamboura, Mr (UDEAO Sec.-Gen.), 79–80
Tanzania, 77n, 78–9n
tariffs: 45, 49, 50, 72, 80, 178; external, 50, 61, 72–3, 144; general, 73, 76; minimum, 73; harmonised 'disarmament', 10, 11, 61, 182; protection, 32; rates, 50
tax: harmonisation, 55; collection, 62; on consumption, 62
tea, 181

telecommunications, 122, 157; and ECOWAS, 127, 128, 130, 140
Tema refinery (Ghana), 134
textile industry, 20, 97, 134, 138
timber, 181
tin, 181
tobacco, 20, 138, 181
Togo: 3, 38, 39, 84, 91, 143; industry, 20; import duties, 59n; and CEAO, 80, 83; and UMOA, 86; and Benin, 88; and Entente, 93, 94, 99n; and R. Niger Commission, 121; and WACH, 123; export and imports, 174, 175; and Ghana, 190
Touré, Pres. Sekou, of Guinea, 107, 108, 109n
trade: creation, 47-8; barriers, 44; diversion, 43, 46-8, 59; integration, 10; intra-African, 157; intra-UDEAO, 59, 78, 80; intra-Entente, 53; intra-Ghana-Entente, 96; intra-ECOWAS, 61; intra-West African regional (zonal), 28, 31, 37, 41, 48, 51, 52, 53, 75, 90, 94, 138, 143, 173
Trans-Africa Highway, 37
Trans-Saharan road (Unity Highway), 37
transistor radios, 112, 134
transport and communications: 3, 31, 35-7, 49-50, 51, 79, 82, 100, 122, 127, 128, 135, 140, 173, 176, 183; Decade in Africa, 173; integrated system in W. Africa, 79, 169
Traoré, Moussa, 107
tree crops, 138
Tubman, Robert, 167
tyre, manufacture, 104

Uganda, 9, 24, 78-9n
under-utilisation of resources, 22, (Ghana) 23
unemployment, 22, 46, 132, 191
Union Douanière entre les Etats de l'Afrique Occidentale (UDEAO), see Customs Union of West Africa States
Union Monétaire Ouest Africaine, see West African Monetary Union
unit of account, W. African, 124
United Kingdom, see Britain
United Nations (UN), 71, 103, 120, 156-61; report on Senegambia, 113-17
UN Committee for Development Planning, 12

UN Conference on Trade and Development (UNCTAD), 122, 155, 156, 185
UN Economic Commission for Africa (ECA); 16, 36, 39, 67, 71, 126, 157; aid to ECOWAS, 155
UNESCO, 157
UN Development Programme (UNDP), 118, 122, 157, 158-60; Development Decade (1960-70), 171
UN Industrial Development Organisation (UNIDO), 106, 122, 156-7
United States: loan to Entente, 99; aid to R. Niger projects, 121; and aliens, 198
United States Agency for International Development (USAID), 100, 106
Upper Volta: 38, 39, 88, 91, 136, 138, 143, 178n; manufacturing, 20, 23; roads, 36, 135; and UDEAO, 72, 74; arrangements with Senegal and Ivory Coast, 77; and CEAO, 80, 81; and BCEAO, 89; and Entente, 93, 94, 95, 96, 97, 99n, 100; and WACH, 123; and Ghana, 101, 190; and R. Niger Commission, 120; and ECOWAS, 134; emigration, 180, 193, 194
urbanisation, 132-3, 186, 196 and n

Venezuela, 53
Viner, J., 43ff
visas: waiving, 169, 192-3, 196; exit, 198

water supplies, 16
weather, see meteorology
welfare, 46, 49, 50, 52, 191
West African Airways Corporation, 67n
West African Clearing House (WACH), 39-40, 68, 70, 71, 123-4, 142-3, 168
West African Cocoa Research Institute, 67n, 173
West African Currency Board, 5, 38, 114, 173
West African Customs Union, (UDAO), 4-5n, 8-9, 31
West African Economic Community (CEAO): 68, 70, 71, 79-85, 126; as danger to ECOWAS, 144-6; as counterbalance to Nigeria, 83
West African Examinations Council, 173
West African Industrial Co-ordination Mission, 36
West African Institute for Oil Palm Research, 173

West African Monetary Union (UMOA), 38, 68, 70, 71, 85–92, 104, 114
wheat, 110, 120, 139, 176; — flour, 138
World Bank, 88, 178
World Health Organisation (WHO), 157

yams, 138

Yaoundé Convention, 33, 34, 82
Yorubas, 3

Zaire, 24
Zambia, 24
Zimbabwe, 24
zinc, 181